THE UNDERGROUND HOUSE BOOK

THE UNDERGROUND HOUSE BOOK

by Stu Campbell

GARDEN WAY PUBLISHING
Charlotte, Vermont 05445

Illustrated by Robert Vogel

Printed in the United States

Third Printing, June 1981

LIBRARY OF CONGRESS CATALOGING IN PUBLICATION DATA

Campbell, Stu.
    The underground house book.

    Bibliography: p.
    Includes index.
    l. Earth sheltered houses. I. Title.
TH4819.E27C35   690'.837   80-14992
ISBN 0-88266-167-1
ISBN 0-88266-166-3 (pbk.)

To Walter Hard Jr.
Whose lightly used editor's pencil
spelled encouragement.

CONTENTS

FOREWORD

The solar age is upon us, and it's here to stay. New technologies have become cost competitive, and the promise of clean, infinitely renewable solar energy is a reality.

Moving hand in hand with advances in energy production, new building forms and methods are being developed to conserve our hard-won energy. Super-insulated shells, passive and active solar collectors, and earth-sheltered designs lead the way in this exciting new field.

As with any technological advancement, it takes a while for mystique to disappear and facts to become clear. "Tech-talk" must be translated into everyday terms, unduly complicated explanations reduced to their original simplicities. "Smoke goes up, water goes down" is the kind of graphic explanation we can all understand, and to put it any less simply is unnecessary.

I've enjoyed working with Stu Campbell on this book because he not only has a working knowledge of building materials and the principles of simple physics, but he can express his ideas in a straightforward manner. The social and economic implications of the new technologies are vast, and Stu wisely flavors this book with the realities of our modern energy dilemmas.

To date, the emphasis in earth-sheltered design has been technological rather than aesthetic, pragmatic rather than poetic. With few exceptions, this preoccupation with practical matters has left architecture behind, resulting in a monotony of plain-vanilla bunkers stuck into hillsides, with the token lamp post and foundation shrubbery proclaiming "HOUSE!"

The great American ranch house facade is banal enough in suburbia, and it's a shame to see it propped up in front of innocent earth-sheltered buildings whose designers haven't the skill or courage to recognize the rich architectural potential in the earth-sheltered concept.

Sculptured earth cover, three-dimensional openings, dynamic structural systems, and the

full spectrum of architectural ingredients —
light, form, space, color, texture, and materials
— are all waiting to be explored.

But most of our architectural leadership is
concerned with "High Art," paying only lip
service to energy-related matters. "High Art"
and practicality are thought to be incompatible,
and the current architectural fashions bear this
out, steadfastly ignoring the hard global issues
of our times.

The need for energy-wise design is urgent.
The solutions can be left to tinkerers and
engineers, but if architecture is to live up to its
rich history, its practitioners must participate
now, as they have from Stonehenge to Chartres,
from the Roman aqueducts to the Verrazano
Bridge, enobling and enriching the human
environment. It's time for the avant-garde
designers to cease their abstract agonies and
apply themselves to the *real* frontier — where
proven technology awaits its long-overdue
artistic companion.

This book recognizes both the practical and
the aesthetic, the general and the specific. It
encourages anyone considering an earth-
sheltered building to examine the full realm of
possibilities before beginning construction.
Above all, it encourages sensitivity in the art of
choosing well, in a given time and place, the
humanistic values that make a house a home.

DON METZ
Lyme, New Hampshire

ACKNOWLEDGMENTS

Without the encyclopedic reference text, *Earth Sheltered Housing Design*, prepared by visionaries at the Underground Space Center at the University of Minnesota, research for this book might have been directionless.

I'm indebted to Roger Griffith, who had the foresight to imagine the project before it began, and has supported it right to the bitter end. My appreciation to Diane Jumper, whose excellent typing I continually messed up, and to Dave Robinson for his tasteful book design.

Thanks are also due photographers Marilyn Makepeace, Ezra Stoller, Ken Basmajian, Robert Homan, Lane, and Phokion Karas. Illustrator Bob Vogel, as always, followed inconsistent directions patiently, and made corrections without complaint. Loaning his visual talents to us as he does is a contribution that can't be measured.

But the real guts of this book come from the designers themselves, especially Andy Davis, Rob Roy, Malcolm Wells, John Barnard, Jeff Sikora, and less directly, the venerable Frank Lloyd Wright. None of these people could have been more cooperative and helpful.

And as for Don Metz, what can I say? Thanks aren't quite adequate.

STU CAMPBELL

1 THOUGHTS

1a. AESTHETICS

I drove across the United States not long ago. After being there for six months, I was relieved to escape California's crowds, gasoline lines, and insanely inflated real estate prices.

I traveled alone in an old friend, an uncomplaining middle-aged Saab, which, despite 87,000 miles of wear and tear, calmly went about the job of transporting me across the country — cheaply, efficiently, in comfort and style. One day I'll live in a house as thoughtfully designed as that car.

I had time to think about the Saab — one of engineering's true gems — and about our highway companions in gray early May, those smoke-belching behemoths, the diesel trucks that form a life support system for America soon to be made extinct by the inevitable petroleum drought. I wondered about the trucks, about their drivers and cargo, and wondered if the gas shortage would rule out transcontinental travel by car.

I had other things to think about besides cars, trucks, and oil. My work puts me in Lake Tahoe, California, for half of each year. I had wanted to buy a home there, but a building moratorium insures that housing demand far outweighs the supply, causing prices to accelerate far beyond my reach. I was disappointed about that. Disillusioned.

While my education in California real estate was going on, Roger Griffith, editor and friend, called me from Vermont and suggested I write a book about underground houses. My first reaction was, "People will *buy* a book about underground houses?" Roger thought so. He'd visited some and had talked to several architects. "I think it's important," he said. I agreed we'd discuss it when I got back East.

On the second day out, the Saab grimly began to eat away at the boredom called Interstate 80, which passes through Wyoming. Flat road. Gray rocks and sagebrush. And the occasional drone of a tractor trailer rig going the other way. This is unfriendly land where hardly anyone lives. I knew I could buy property here — cheap — but wouldn't want to. To me being on the high plains is like visiting a strange planet.

I began to look for signs of people. The rare ranches and homesteads (usually mobile homes) looked naked and vulnerable. Painted metal walls had baked and flaked in the summertime sun. Snow fences and bent antennas testified to the relentless battering of northerly winter winds. Depressing and barely inhabitable country — where acres and acres are needed to support a single head of livestock.

By midday, miles from nowhere, the sun made its presence felt just enough to reveal a backdrop somewhere behind the monotony. From the haze to the north emerged high mesas and semi-barren mountains — with snow. And foothills, hundreds of little rolling humps, tier on tier of them, backing up to a sparse but very definite timberline. All the south-facing hillsides pointed at the warm sun, and since the thin snowpack above had started to melt, the slopes were green in the spring runoff. Their color clashed with the pallid bottomland far from the hills.

The possibility of year-round wells at the base of these green slopes brought more focus to the picture. I might have seen a busy airport built on the flats near communal wells. I thought I could see underground waterlines burrowing up the hills toward large panes of glass that enclosed snug houses dug into the hillsides. When I looked more closely I could see chimneys coming out of the ground, next to antennas, and wind generators, and solar collectors.

Space-age Hobbit holes! From the chimneys there was smoke from wood fires, fueled by the

trees higher up. There were gardens and orchards and vineyards over and around and among the many windows in the landscape. Sun streamed through the glass, heating living space within. There were roads with gasless vehicles and people scurrying about business as usual. Businessmen commuted by air to Cheyenne, or San Francisco, or a little further to Tokyo.

Things were looking brighter, but when I put on sunglasses — of course — I saw the underground village in my mind's eye wasn't there yet. Only the sunny green hillsides remained. Waiting.

Weeks later, as I gnawed at the lean body of literature dealing with subterranean building, I was astonished to read a passage from Malcolm Wells, perhaps the world's best-known architect for underground living. His vision was a lot like mine:

> At the moment my dream job is this series of little communities on the lower slopes of some mesas in a huge western tract of which 95 percent will be left wild. Cars will be virtually barred, food will be grown on certain flood-plains, and all other energy will be gathered from the sun and wind. The nicest part of all is that because conventional zoning will be replaced here by ecological zoning, many of the units can be built tiny enough to drive their costs down in the range of the average, low-to-middle-income buyer.[1]

I also saw a *New York Times* piece about Don Metz, an architect in Lyme, New Hampshire. I liked the photographs of underground houses he designed. I found out more about him, then called him up.

"It seems to me that underground houses are only for rich people who can afford to experiment," I said over the telephone.

"I don't believe that," he said. "We're building a house right now that's designed to disprove that idea." He was talking about Earthtech 5.

I wanted to see, so I drove to Lyme a few days later. I found the simple plywood sign that said "Metz," followed the long gravel driveway up the south-facing hillside that recalled southwestern Wyoming, and parked next to a white Saab even older than mine. This was the site of Baldtop Dugout, which I'd read about.

Metz came out of a separate above-ground studio/garage in work clothes, and his handshake spoke of calluses formed by more than once-in-a-while contact with a hammer and trowel. He told me later he was an architect mostly because he liked to make things.

---

1. Malcolm Wells, "To Build Without Destroying the Earth," *The Use of Earth Covered Buildings*, p. 212.

Marilyn Makepeace

Baldtop Dugout, Lyme, New Hampshire.

Inside the studio I told him as much about myself as a few minutes would allow. We looked at several sets of underground house plans he was working on. One house was being built in Dallas, one in Maine, and a third, Earthtech 5, was under construction just down the road in Lyme.

I asked, "Could you design a 1500-square-foot house — about the size of the average new home being started in the United States in 1979 — for 10 percent below the cost of a comparable above-ground house? And could it run on half the energy of the conventional house?"

His answer was quick. "No."

A properly built underground house, he explained, would probably cost 10 percent *more*. But the energy savings should be *more* than 50 percent — possibly 75 percent. The extra construction costs would be paid back very quickly.

Together we walked to the crest of the hill and down the garden-like entryway into Baldtop Dugout, Don Metz's underground, owner-built home. As I stepped through the door, any negative feelings I ever had about living below the earth's surface were immediately turned around. I was so comfortably disoriented, in fact, I had to ask, "Which way is south?"

"That glass wall is the south side of the house," said Metz.

In each of the mysteriously curved spaces of the main level — spaces that seemed to start and end at no place in particular — I felt sheltered by the heavily timbered roof, yet still in the open somehow. We came from the garden, milled around in the living room discussing the family cat and the brickwork in the fireplace, ate lunch in the dining room (or was that part of the kitchen — just off the bedroom with the unmade bed?), then walked down a spiral staircase, through a glass door into the garden again. We'd toured the house, but it was as though we'd never been "inside."

Stu Campbell

Don Metz, AIA

From here we drove to Earthtech 5, a passively heated underground home that's glazed on both the south and west walls. It's the fifth in the Earthtech series of Metz's design. Only two-thirds completed, it was already attractive. Not so large as Baldtop Dugout maybe, and surely not so fanciful, but beautiful in its straightforward simplicity and tasteful use of space, sun, and thermal mass.

I was sold. As the afternoon wore on, Don Metz and I sat in the car mulling over some of the basic things people should know about underground housing. We agreed to talk regularly, and began to formulate plans for designing Earthtech 6.

That's how this book was conceived.

1b. ATTITUDES

". . . But isn't it cold and dark? Aren't underground houses damp? What about condensation? How do you get out if it collapses? Doesn't the air get stale?"

Questions like these — and they always come up — reflect public attitudes about underground living. Many are openly hostile to the idea. "No way!" they say. "I'm no mole!"

Others express genuine concern for the problems of waterproofing (see 5g), ventilation (7h), roof structure (6aa), and natural lighting (8d). There are totally satisfactory — even pleasing — architectural and mechanical answers to each of these questions, of course. In fact, technical solutions have been around for some time. It's just that most of us haven't been educated yet. We need to be shown.

Any number of things explain our negative

feelings. A lot of it is cultural. Heaven is always up; Hell is down. When we die, we get "planted" in the earth. One of man's worst nightmares is that of being "buried alive." Almost as bad is the thought of wandering lost in dark, endless catacombs.

Accounts of horrible mining accidents — where man-made supports have failed — rivet our eyes to television news broadcasts. We like to think our prehistoric ancestors were smart enough to move *out* of their caves, gradually learning to build shelters from sticks and skins, stones and sod, mud, boards, and glass. Later still, we discovered asphalt roofing, vinyl siding, and fiberglass insulation. There are now those who say we weren't so smart after all.

1c. Basement Apartments

Our attitudes are also based on past experiences with leaky cellars, moldy crawl spaces, foul sewers, and dingily lit subway stations — homes for rats, thieves, and perverts. Any experimental underground houses we've seen looked like pill boxes or gun emplacements. And we associate downstairs apartments — attractive as they might be in terms of energy savings — with "bargain-basement living."

The point is that we've been very one-sided in our thinking. Because Mother Earth offered so much habitable surface area, man, until lately, hasn't had to move to more hostile areas where he might have to "dig in." There's been so much attractive land, and such an abundance of energy, that we haven't had to face up to the problems of using space without destroying it. When we got a little crowded, the obvious solution was to build up. The results: scenes like the Manhattan skyline.

In spite of our squandering much of the best surface land available, in spite of archeological evidence of ancient peoples who "hunkered down" in the earth in comfortable communities that stayed cool in summer yet received the low angle of the wintertime sun, in spite of the existence of a newly built underground bookstore at the University of Minnesota so energy-efficient that its lights and people generate enough heat to maintain a 70° F. indoor temperature until the air outdoors gets down to −20° F. (at which point the active solar heating system kicks in), and in spite of the losing battle that our most advanced above-ground dwellings are waging with heat loss and overconsumption of fossil fuels, we continue to think "up."

1d. Bomb Shelters

In recent history, when we *have* thought about going underground, it's been for strange reasons. In the 1950s, we saw covered and windowless holes in the earth as places that offered protection from atomic fallout after a nuclear attack. These "bomb shelters" were dreary crypts usually, with enough stored food and fresh water to sustain a family for however long it might take for the radiation danger to pass. These would not have been pleasant places to spend much time.

In the early 1960s, Jay Swayze, a Texas designer, built an elaborate and much-publicized underground vault complete with artificial views, lighting that changed throughout the course of the "day," mechanical ventilation and homey features like a tiny fan to ruffle curtains when you raised a "window" opening to nowhere. All power for the luxurious shelter/ home was to be supplied by a large generator.

Not long ago the Swayze-built house, which cost well over $100,000 to construct, was sold for less than $50,000. The new owner remarked that he didn't relish the idea of living underground, but could never have found a house of such quality in such a good neighborhood for such a low price — above

ground. Swayze also built an underground exhibit for the 1964 New York World's Fair. It was received with only moderate interest.

Now in the early '80s as the Swedes, Russians, Japanese, and French build below ground more and more frequently, American designers still find themselves running public relations campaigns to promote the idea. They wisely avoid the word "underground" with all of its negative connotations, and instead refer to their below-grade structures as "geotectures," "terratectures," or "earthen homes." The most popular euphemism is "earth-sheltered housing."

The challenge these architects face is this: It's not enough to show us their houses and have us say, "This looks comfortable; this is safe; this is not so expensive," or even, "This is energy-efficient." We have to walk in and say, "Look — this is beautiful!" Only then will we believe.

1e. ARGUMENTS

By 1985, 30 percent of our residential space and 40 percent of all commercial buildings will have been constructed after 1974. That's what the Energy Research and Development Administration (ERDA) tells us, anyway. How

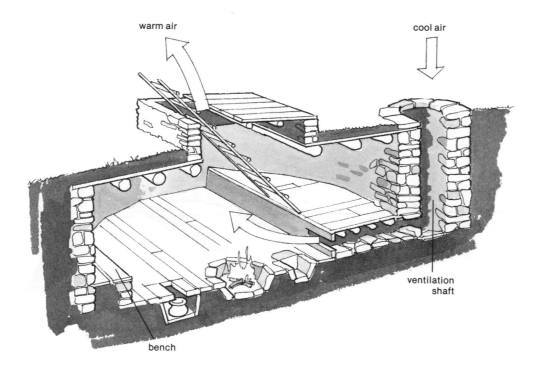

warm air

cool air

ventilation
shaft

bench

Underground "kivas" are still preserved in Mesa Verde National Park. They're the only archeological evidence of underliving on the North American continent. At first, they were used for religious purposes. Later, the ingenious ventilation system made them suitable for habitation.

much of this new space will be underground is hard to predict.

Proponents of underliving would have us believe that in-earth structures will constitute as much as 30 percent of the housing market by the mid-1980s — an optimistic projection, it seems. The *Wall Street Journal* estimates that as few as 3000 earth-sheltered homes were built by the end of 1979. Clearly, anyone who started a subterranean house before 1980 must be regarded as a pioneer.

To be brutally realistic, those who plunge into the underground housing field do so with little precedent. Descriptions of subsurface dwellings discovered in North Africa, Turkey, Ethiopia, China, and Israel are impressive, but

on this continent, underground history is limited to the ceremonial "kivas" built by American Indian tribes in the Southwest. These are still preserved in Arizona's Mesa Verde National Park.

Our Midwestern settlers built their first houses out of sod because no other building material was handy. But this particular type of earth shelter must have been miserably cold and drafty. Sod houses were above ground, in most cases, exposed to the ceaseless prairie wind. In spite of what we think, soil, in itself, is poor insulation (7a and 7b), and heating experts calculate that a typical sod house wall had an R-factor of about 1, meaning that it was about as effective as thin plywood.

Even though the concept is new to this part of the world, arguments for the use of underground space are hard to resist. They focus mostly on the preservation of surface space (2a); privacy (2k); improved, vibration-free acoustics (2l); structural stability (6h); low maintenance (3h); and most important of all, low energy consumption (3f and Appendix 3).

It's no accident that much of the serious research and persuasive hoopla for earth-covered buildings comes out of the Underground Space Center at the University of Minnesota. A staggering 33.3 percent of all the energy used in that northern state for domestic purposes goes directly into heating living space. To Director Ray Sterling and others who work at the Minneapolis center, underground construction is a matter of urgent economic necessity.

Earth Sheltered Housing, a book originally prepared by the Underground Space Center for the Minnesota Energy Agency, had a modest first printing of 4000 copies in 1978. But it was practically the only title in the field, and by mid-1979 the book had reportedly sold close to 70,000 copies — mostly by word-of-mouth — and had been reprinted by Van Nostrand Reinhold in New York.

Any builder or Realtor who suggests there's no interest out there has *his* head buried in the sand. To its surprise, perhaps, the Underground Space Center has a tiger by the tail. Together with the American Underground Space Association, they sponsor public conferences called "Going Under to Stay on Top." Held all over the country, these public seminars attract overwhelming attendance. Jay Swayze, who could generate little underground interest at the '64 World's Fair, must be amazed.

Maybe the prophecy that 30 percent of us will be living underground by 1985 is *not* so far-fetched.

1f. ARCHITECTS

Ken Labs, a young architect from a firm called Undercurrents in New Haven, Connecticut, is an energetic spokesman for the underground movement. In this final, polluted, and shortage-ridden quarter of the twentieth century, his promotional rhetoric is as unassailable as motherhood and apple pie. Underground architecture, he says, accomplishes two things — energy efficiency and the preservation of environmental quality. Who'd want to argue with that?

Frank Lloyd Wright's Solar Hemicycle was designed in 1943, but the concept was borrowed from a similar Wright building — a boathouse in Madison, Wisconsin built before the turn of the century. With its earth berm, passive solar heating and cantilevered roof, it's not far from state-of-the-art underground homes of the 1980s.

Labs's energy-efficiency claim is reinforced again and again by hard statistical data from engineers in Minnesota, Georgia, Texas, and at MIT who have been monitoring several experimental houses. The idea of preserving environmental quality may be more difficult to prove, but it echoes architecture's most venerable legend, Frank Lloyd Wright, and his fictional counterpart, Howard Rourke, hero of Ayn Rand's *The Fountainhead.* Both believed that architecture should disturb the environment as little as possible; that a building should blend, and all but disappear into its site.

1g. Frank Lloyd Wright

Wright, with an incredible record of innovative designs to his credit — ideas we take for granted today — has almost never been associated with underground architecture. Yet in 1943 he designed a home for Herbert and Katherine Jacobs that was built in Madison, Wisconsin. He named it the Solar Hemicycle, "suitable," he said, "for any spot in the country where there is good drainage, for the house creates its own site and its own view." Mr. Jacobs describes a meeting with Wright, when they first went over the sketches.

> Reminding us that we had been concerned about the cold . . . winds of winter, Wright pointed to the semi-circular shape of the house, which faced a half circle of garden sunk four feet below floor level — the floor level itself being a foot and a half below grade level. Facing the sunken garden, behind a narrow terrace, a solid band of glass and windows, forty-eight feet long, rose fourteen feet to the overhanging roof. Behind the house, at the north side of the structure, a slope of dirt reached almost to the narrow band of windows, about a foot high, which circled the entire rear of the house.

> With the big windows at the front, the low-lying winter sun would come way into the house and help to warm it, but the high summer sun would be cut off by the roof overhang, Wright said. The other feature to counter the cold weather, he declared, was the creation of an airfoil by streamlining in place.

> "The sunken garden in front of the hemicycle acts to form a ball of dead air, and the long slope of dirt against the back wall is necessary so that the wind will blow over the house instead of against it," Wright said. "When it is finished, you can stand on your front terrace in a strong wind and light your pipe without any trouble. With little wind blowing against it, and with the mass of dirt at your back, your heating costs will be very low." [2]

Wright had used this same concept — including the passive solar heating — in a design for another building fifty years before! His early plans, which included berming and overhangs that considered the angle of the sun's rays, are very close to today's state-of-the-art designs.

1h. Malcolm Wells

Malcolm Wells, quoted already, believes that otherwise dead and undesirable building lots can be brought back to life by earth-sheltering. "Ugliness can be swept under a carpet of earth," he says. His famous design, Solaria (9c), completed in 1976, was built for the Robert Homans, near Philadelphia. Wells, as usual, tells his own story best.

---

2. Herbert and Katherine Jacobs, *Building with Frank Lloyd Wright*, p. 83.

Rear entryway to Malcolm Wells's Solaria

All through the winter of '77 we waited for word from the Homans. January. February. March . . . Finally Bob visited us, and we all said, "Well?"

"We didn't use the auxiliary heat at *all.*"

So we had our first real proof that under record conditions, solar heat and earth cover *is* a powerful combination, even in the northeastern part of the United States.[3]

As he pleads the case for underliving, Malcolm Wells suggests secret longing, on man's part, to go back to the primeval cave, or further still, to the womb.

Womb-like or not, underground construction usually demands no special trade skills. Any competent builder should be able to do an adequate job with readily available materials. Still, because the house must integrate closely with its site, consulting with a qualified architect should be a high priority.

But not just any architect. It's best to find someone familiar with the intricacies of underground building. Standard plans, like the countless designs that now exist for conventional houses, will be available in time, no doubt, but they're rare at the moment. Even

---

3. Malcolm Wells, *Underground Designs,* p. 10.

Malcolm Wells's New Jersey office lies beneath what everyone else considered a "bad" building lot. Just twenty feet from a major freeway, it contains "silent, sunlit rooms." The roof curbs, made of scrap lumber, rotted away while the roofscaping established itself.

if earth-sheltered design can be reduced to formulas, each building site will have its unique set of problems, which good architecture should deal with.

It's critical to remember that building below the surface is not the same as building above — in the atmosphere. It means working in an altogether different medium, as we'll see.

2 LAND

2a. SITE PLANNING

The Japanese have long believed that land is its own architecture. Often, but not always, they've tampered with it more subtly than we — without bulldozers and earth movers. Their best buildings harmonize with their settings, sitting camouflaged among rocks and greenery.

We, on the other hand, have grown fond of using the brute force we call Machinery to change the face of the earth. When a house fails to fit a site, we change the site — rarely the house itself. Perhaps it's because, unlike the Japanese, we've had so *much* space. Maybe it's

because earth-moving has been cheap. Or maybe it's because we haven't known any better.

The lay of the land needs to be studied closely before deciding which type of earth-sheltered housing best suits a particular lot (see 6b). The quality of the site may also determine the number of people who eventually live there. Present suburban thinking usually allows a maximum of three to four dwellings per acre, and zoning ordinances (2i) in many places find *that* too crowded.

Planners for the underculture see it differently. They forecast that well laid-out sites

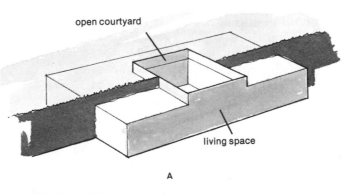

A flat site might accommodate a completely recessed house (A), or one that's semi-recessed (B).

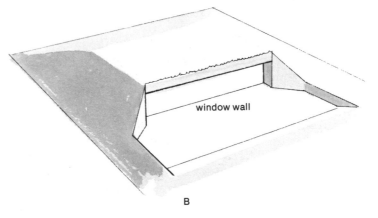

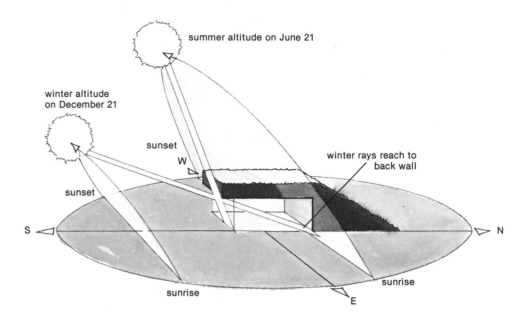

summer altitude on June 21

winter altitude
on December 21

sunset

W

winter rays reach to
back wall

sunset

S

N

sunrise

sunrise

E

Designing for passive solar heating involves careful site selection, an understanding of solar altitudes, and correct orientation on the lot.

can accommodate as many as seven earth-sheltered units on an acre, provide privacy for everyone, and still leave 85 percent of the land open for gardening and natural vegetation.

2b. Exposure

Forward-looking Realtors also predict that south-facing lots will soon sell at premium prices. In some parts of the country, where buyers are more energy conscious, they already are. Shaded land that points north will sell more slowly, and prices on such property, which in time may be suitable only for woodlots, will remain relatively stable.

Ideal sites for underground homes are sunny most of the day, so the house can be heated by the sun in winter. In summer, solar heating will have to be reduced, perhaps by a cantilevered roof that overhangs and shades any windows that face south (8d).

Some of the best current designs allow for "passive" solar heating (7c), where warmth is stored in masonry or in sealed water tanks to be released at night when there's no more input from the sun. "Active" solar systems, which include shiny collectors, heat-transfer mechanisms, and storage elements, may also be considered.

Anyone looking for a suitable underground site should carry a pocket compass and refer to it often. Southern exposure, even more than view, should be the number-one consideration. The red end of the compass needle, of course, will point to magnetic north; the white end will point to south, which may be several degrees away from true south. When the house is actually oriented into the site, its windowed side should face close to true south.

The magnetic variation — between true north and magnetic north — is easy enough to find. It appears on any Geodetic Survey map or aviation chart, and a call to the nearest airport — even a tiny one — should tell you the variation and whether it's east or west.

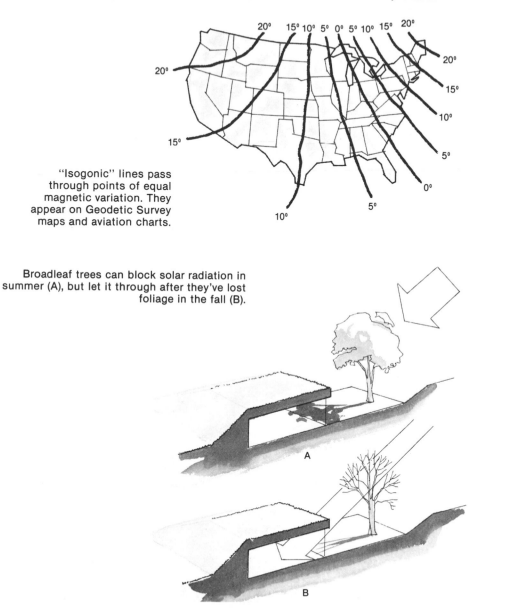

Easterly variation   Westerly variation

"Isogonic" lines pass through points of equal magnetic variation. They appear on Geodetic Survey maps and aviation charts.

Broadleaf trees can block solar radiation in summer (A), but let it through after they've lost foliage in the fall (B).

Remember the pilots' buzz phrase, "East is least, but west is best." If the local variation is "16 degrees west," true north is 16 degrees *plus* — clockwise of where the needle points. If it's "20 degrees east" true north is found by *subtracting* 20 from magnetic north — counter-clockwise on the compass.

The perfect exposure for a window meant to collect solar radiation is 15 degrees west of true south, but 20 degrees to either side of this point is still excellent. Fine passively heated solar homes have been built on lots that do not face directly south, but designing them is more complicated.

Look for mountains or hills that block low incoming rays from the sun during the winter. Evergreen trees, which keep their foliage even in the cold season, can obstruct solar radiation too, and may have to be sacrificed. Broadleaf trees should let sunlight through once they lose their leaves in the fall.

2c. Topography

Picking a site for an earth-sheltered home is like selecting a Boy Scout's campsite. Surface water (5a) must be taken into account — even if it's not there at the moment. The Scout who pitches his tent in a low spot, at the foot of a gully or small flood plain, or even on ground with a high water table (5c), may learn the hard way — waking up in a drenched sleeping bag.

A house is naturally more permanent than a tent, and one that's dug into a poorly chosen site may end up in a state of perpetual wetness.

The shape of land — its topography — affects more than the surface runoff. It also influences wind patterns, vegetation, and shading, and limits the way roads and overbuildings can be planned.

In winter, prevailing winds blow from the northwest. This side of the house requires the most protection — from the hillside itself (if there is one), from an artificial mound of earth

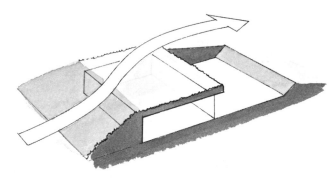

Malcolm Wells said that a house sited with its back to the wind "pulls a blanket of warm earth over its shoulder."

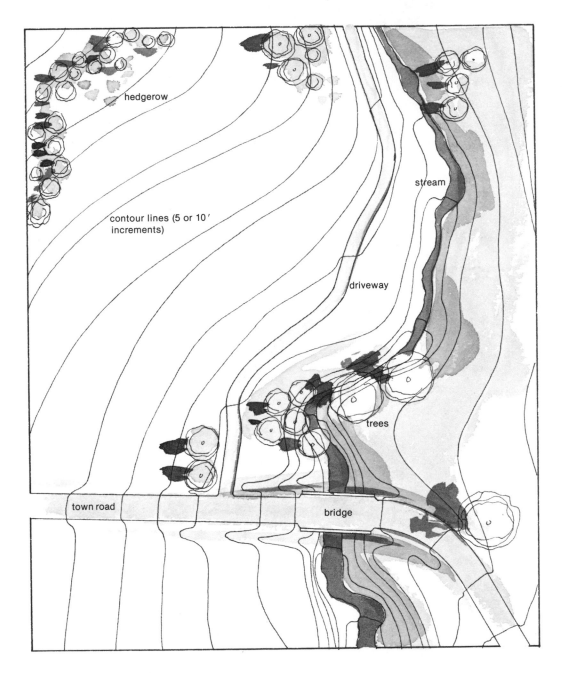

hedgerow

stream

contour lines (5 or 10' increments)

driveway

trees

town road

bridge

A contour map of a real lot helps in decisions about how and where to site the house.

(called a "berm"), and possibly from a windbreak of shrubs and trees. Any windows on the northern side of the house will produce heat loss (8d), but very small ones may be necessary for cross-ventilation during hot seasons and for humidity control (7h).

The very best building sites will slope gradually away from severe winter winds — toward the warmth of the sun.

2d. Soil Study

Swamps and areas of permafrost are out of the running as potential sites for underground construction. So are geological faults and earthquake zones — though a well-built, earth-sheltered house is probably safer than a surface building in a place with a history of light seismic activity. Minor earth tremors should have little effect on a strong underground structure.

The soil on the site not only has a bearing on the ability of the house to stay put without settling (4b, 6k) and on the septic and drainage systems (5f); it also influences the amount of stress that's brought to bear against the walls (4d). A high water table, where saturated soil sits close to the surface, can cause damp feet, and waterproofing problems that are more severe than normal. So the water-holding capacity of the ground should be scrutinized closely.

Before making a final decision about a site, examine soil samples, take test borings to learn the depth of soil in various parts of the lot, determine the ground water depth and movement patterns, and do percolation tests. At 1980 prices, all of this might cost upwards of $700.

2e. Neighbors

How is the possibility of your living there going to affect the neighbors? How will they affect you? Is your excavation somehow going to undermine someone else's foundation or footings?

An underground house will cast little or no shadow, but adjacent homes may interfere with *your* sunlight (2j). If you live too close to the people next door, you may not only live in their shadow, you may feel you're being looked down upon as well — literally.

Excavating and pouring concrete walls underground may change the flow of subterranean water, and affect the homesites of adjoining landowners. Investigate.

Soil studies may show "ledge" — bedrock that's too close to the surface. Ledge often

means unacceptable leaching conditions for sewage waste disposal. Large holes *can* be dug, and sand and gravel hauled into them to absorb liquids already treated in the septic tank, which in itself requires a minimum of four feet of diggable earth to be buried.

But if the bedrock is impervious, even this artificial leach field becomes a catch basin with no outlet, a condition the county's soil conservationist will look upon with disapproval. If too little attention is given to proper sewage disposal — and there are no environmental laws to insure it — the health of the whole neighborhood may be endangered.

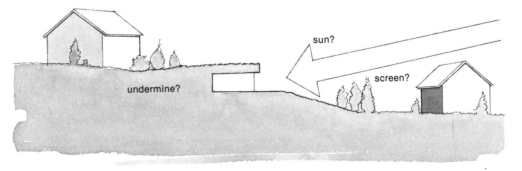

Is there any way a neighboring building can obstruct your sun rights? Is your excavation going to undermine someone else's foundation? Will you need a screen of vegetation to prevent neighbors from peering through your front window?

Ledge can also mean expensive drilling and blasting to create underground space for the house (4f). Shock waves from blasting have been known to crack foundations nearby. Ledge might also call for a man-made hill to cover the roof of the house, which in turn, calls for truckload upon truckload of fill. Dollar signs again.

Shallow soil at the site presents one final problem. Building the berm and covering the roof may take twice as much soil as you have on hand. Soil can always be moved, even trucked in if worse comes to worst, but there must be enough space for the storage and transportation of dirt from one place to another, without encroaching on the neighbors.

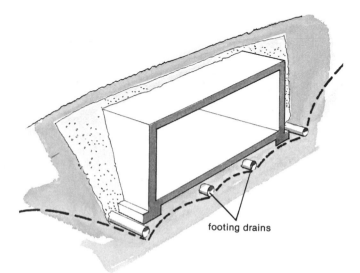

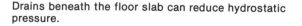

Drains beneath the floor slab can reduce hydrostatic pressure.

2f. Soil Drainage

In short, an underground building site *must* have adequate drainage, whether it sits on ledge, hardpan, clay, or any of the more favorable soil types (4a). Excessive soil moisture increases the chances of a damp living area and causes unnecessary hydrostatic pressures to build against walls (5e). It also conducts more heat away from the sides of the house (5e). These are just a few reasons why proper backfilling

techniques and the correct installation of footing drains (5f) are critical.

The textbook site will have adequate space, good digging, superior drainage, a gentle southerly slope with bedrock well below the surface, and it will be located in a seismically quiet area with a low water table. Finding a spot with all these qualities won't be easy.

To bring natural light into all parts of the house, the earth berm may have to be penetrated.

2g. BUILDING CODES AND ZONING

Underground building is still a relatively new idea. In the eyes of those who write building codes and zoning ordinances — the local laws stating exactly how and where a house may be constructed — there are few restrictions on the use of shallow underground space. Those who *do* it first will form public attitudes, and revise the rules and regulations surrounding terraculture.

Right now there are as many as 1700 recognizably different building codes in the United States. Very individualized differences in public policies can be found from one town to the next. For example, Town A may have no restraints whatever, while Town B, three miles away, may show prejudice against partially completed homes through an ordinance that bans "basement houses."

In this case, those who have built homes a little at a time, by pouring a foundation, sealing it with a waterproofed deck, and moving into the cellar to wait for more capital, have given underground housing a bad name among neighbors, lenders, and building inspectors. Laws like this may be difficult to buck in some places.

2h. Typical Codes

Usually it's easier to design a house to conform to local standards than it is to seek a variance — a process that often results in frustration and delay. Acceptable window space and natural lighting requirements often prove to be the biggest hurdles for the underground house designer, but there are plenty of ways around them (6b).

Typical codes state that each habitable room must have a fire-escape route and natural lighting. Some still require natural ventilation in each room, although mechanical ventilation is permitted in most places. Federal Minimum Property Standards (MPS), after which most local codes are modeled, insist that rooms other than bathrooms and storage areas have windows

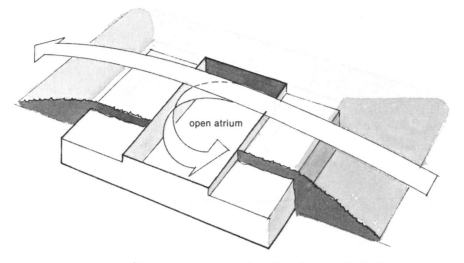

open atrium

In a way, an open courtyard is a huge skylight. Sometimes it's the single natural light and ventilation source for the house. At other times it can be a source of wind turbulence.

greater in area than 10 percent of the floor space.

As heat loss becomes a more and more important factor in building design, it's anticipated that glazing requirements will soon be reduced from 10 percent to 8 percent. As it stands now, in many codes, wherever there is no mechanical ventilation, a room must have opening windows equivalent to at least 5 percent of the floor space. These kinds of details are things with which an architect should, and no doubt will, be familiar.

The uppermost surface of a building is known as "roofing material" in many building codes. Here the type and quality of roofing is clearly specified — ostensibly for everyone's protection. But this specification may present another potential snag to someone wishing to build below ground.

Earth may *not* yet be considered a suitable roofing material, and may require special approval. Some architects have had success by showing detailed plans for a sound roof structure (6y, 6aa), and by arguing that the intended house was to be "built at grade" rather than "underground."

An architect might solve the fire-escape problem by planning built-in storage trunks that can double as steps to a window in an emergency. Rope ladders leading through opening skylights (8b) offer another possibility — which may or may not be locally acceptable.

The natural lighting question may be answered by skylights, too, by lining up rooms in motel-like regiments so each has an exterior wall with glass, or by penetrating either the center or the ground near the periphery of the structure so light can flow into each room (6b).

2i. Zoning Ordinances

There are other things to worry about as you look at possible building sites. Local ordinances may demand a minimum lot size. Very specific septic tank requirements can rule out an otherwise appropriate location. Road frontage restrictions, height limitations, and fire-lane regulations may also apply.

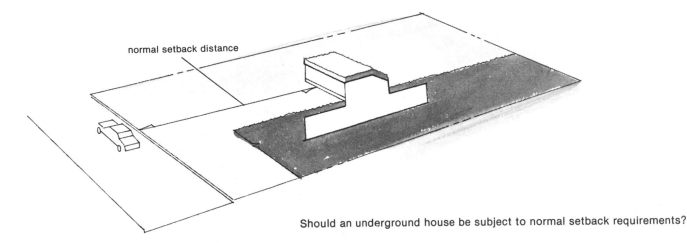

normal setback distance

Should an underground house be subject to normal setback requirements?

In most towns "set-back" requirements say you may build no closer than a prescribed distance from your own property line without breaking the law. As earth-sheltered houses become more and more prevalent, existing set-back ordinances will become increasingly wasteful, and obsolete. And the inevitable question will be raised, Shouldn't an underground house be subject to different set-back guidelines? This is a technical and legal gray area today, but local ordinances are sure to be challenged soon and new policies established.

2j. Predictions

Apart from the normal set-back requirements already dictated by zoning, the question of solar blockage (2e) will become a heated issue long before 1985. Resolutions will be forced by poorly planned underground houses that have their sunlight cut off by neighboring buildings. "Sun rights" is a phrase not yet part of the vocabulary of most zoning boards, but it *will* be soon. And its ultimate definition may come out of the courts.

In addition, building codes and zoning ordinances will have to address three other questions right away. First, waterproofing standards need to be set (5g). Second, we need better structural-capacity guidelines (6h). Finally, we must have inspectors and appraisers who are educated in the field of subterranean design.

As time goes on changes in codes will accommodate earth-sheltered housing. Wisconsin's new and progressive building code, which became effective on January 1, 1980, is probably the first state-wide policy in the country to acknowledge the unique characteristics of underground building. Others are sure to follow as land availability dwindles and demand intensifies.

Legal authorities in 1980 disagreed on the form the evolution in building ordinances will

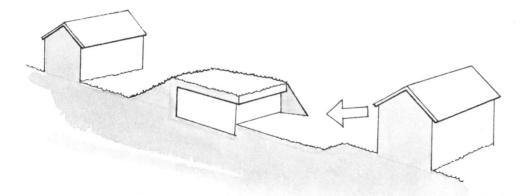

An earth-sheltered house might relate poorly to overhomes in a crowded residential zone. There would be shading problems and line-of-sight problems. But if all the houses were underground. . . .

take: the modification of existing laws or a total redefinition of regulations governing structures below grade. Maybe the near future will bring us "dual zoning," where amended ordinances recognize the feasibility of a higher population density below ground (2a). Either way, there will be critical test cases in the next decade.

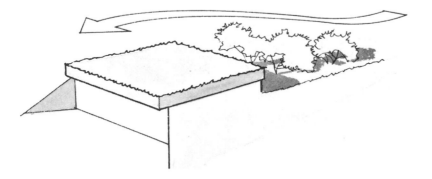

Thoughtful plantings on the north and/or windward sides of a house can act as a windbreak and help insure privacy.

2k. PRIVACY

If a building site is not well planned, lack of privacy can be a problem. At some point, the close proximity of underground dwellings combined with the needs for private space, gardening room, and the importance of sunlight, are bound to produce some painful legal ramifications. At this stage what these will be is unclear.

In the meantime, earth-sheltered houses, like solar homes, are curiosities. Once word gets out they exist, everyone wants a look. People who live in them report being startled by uninvited or unannounced visitors who unexpectedly appear at a window.

Human traffic must be gently guided toward, into, around, or away from different parts of the house. Entryways must be clearly defined. Bedroom windows must be screened from probing eyes. And rooftops must be protected from unwitting heavy equipment operators who either don't know or forget they're there.

Some urban architects suggest the need for "approach control" — electronic warnings or closed circuit television cameras to make an unexpected arrival less abrupt.

Other designers who prefer more "organic"

architecture, see no need for sophisticated hardware. Instead, they say, the important privacy factor should be taken into account at the very beginning of the site plan, and again later in the sloping of earth berms (4e), roof plantings (4i), retaining walls (6y), and landscaping (4j).

21. ACOUSTICS

Natural soundproofing is a major fringe benefit of living in the ground. Malcolm Wells's subterranean office in New Jersey is, as he puts it, "light years away from the freeway twenty feet from its wall."

Inside there are "silent, sunlit rooms." And in Armington, Illinois, when a herd of cows in a pasture next to Andy Davis's underground "Cave" (9a) broke loose and stampeded across his roof, Mrs. Davis, busy inside, never knew what happened.

The sound-dampening qualities of soil seem to offer the perfect cure for noise pollution. I know from experience that nothing could feel more secure than sleeping in the lower level of Don Metz's Baldtop Dugout. A new generation of underlivers is finding that trucks can rumble past and planes can roar overhead, hardly noticed. Purring refrigerators and burbling water heaters become the big sounds — deafening all of a sudden.

Malcolm Wells is right — about "unattractive" sites, that is. An inexpensive piece of land next to a noisy schoolyard may be just the thing. Or a sunny lot at the end of a loud airport runway could become the quietest home you'd ever want. If the home were made to suit the site, hardly anyone would know you were there. And vice-versa.

3 MONEY

3a. FINANCING

It may be the same old story. Financing an underground home may be like getting a building permit — frustrating. Again, newness will be your strongest adversary. Bankers are people after all, with attitudes just like anyone else. They're suspicious. And skeptical of what they don't understand. They don't know enough about earth-sheltered housing yet.

Even though some people have been discouraged by fruitless attempts to find mortgage money, there are few, if any, federal regulations or lending restrictions against underground houses. The Federal Housing Administration (FHA) has not allowed much yet, but its officers approve slab-on-grade construction consistently, and they're rapidly being educated about the advantages of dropping a slab several feet *below* grade.

FmHA, the Farmers Home Administration, is showing ever-increasing interest in energy-efficient housing, and is already knowledgeable about the favorable life-cycle costs in earth-covered buildings (3h and Appendix 3). All indications are that the government is falling into the ranks behind underground building, and loan money from the federal agencies should be flowing soon. Local banks will follow suit — as they normally do.

3b. Default Data

Many bankers, at this point at least, are scratching their heads wondering about the resale value of underground homes. The few mortgages that are being written are done so "cautiously." Bankers worry, as always, about a borrower defaulting on a loan, and about their ability to unload the house on another buyer to regain their investment.

There is no evidence *yet* that potential defaulters are any more likely to live below ground than they are above the surface. Hesitation on the part of loan officers results as much from lack of statistics about defaults by underground loanees as from questions about the marketability of earth-sheltered dwellings. As long as they know next to nothing about subterranean design possibilities, lenders will continue to drag their feet.

At the moment, there's a vicious cycle going on. Speculative contractors need capital at attractive interest rates if the underground construction business is ever to get off the ground. More underground speculation would

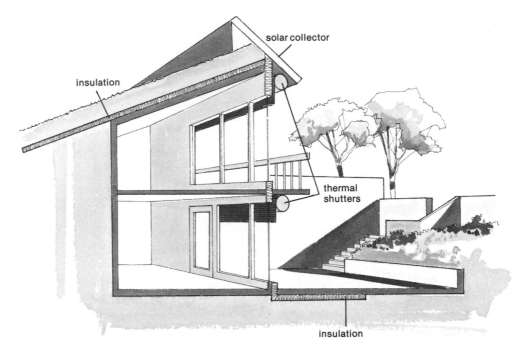

insulation

solar collector

thermal shutters

insulation

Take pretty pictures to the bank. Emphasize lighting, energy efficiency, and resale potential.

produce more favorable data to help lending institutions make decisions about loan applications. But little seems to be happening at either end.

Demonstration projects, like the one subsidized with $500,000 of state funds in Minnesota, can help break the financial ice. When one pilot house outside Minneapolis was opened for public inspection on a drizzly Saturday morning in March of 1979, hundreds turned out to see the earth-sheltered structure. Hundreds more never made it because of the discouraging traffic jam. Before lunch there were several serious offers from prospective buyers.

Mary Tingerthal, of the Minnesota Housing Finance Agency, which helped put the project together, says that already there is far less resistance to underground mortgages in Minnesota than might be expected.

3c. Market Value

Ms. Tingerthal suggests, "Take your lender to lunch . . . and get in to see the highest officer of the bank you can." He may be freer thinking, more visionary, and less conservative than a low-ranking loan officer, she points out.

But he'll almost certainly question the resale worth of the property.

Any lender will want assurance that the design conforms to local building codes. If the home is too different from other houses in a neighborhood, that may affect its market value. On the one hand, novelty can make an underground home attractive for resale, on the other, lack of visible "presence" (1f) could detract from its resalability. As you design, avoid making the house so customized and individually tailored that it becomes unappealing to possible owners in the future.

Banks are aware that energy costs, rising as they are, may soon put people into financial binds that eventually lead to default. "Variable-rate mortgages," which will allow monthly payments to fluctuate with the economy and money market, are bound to become more common shortly. These will insulate borrowers somewhat, but it's generally accepted that energy-efficient housing protects people from spiraling costs even more.

Your ace in the hole will be to point out what the banks already know: that energy costs will surely rise faster than incomes in the next few years. "Life-cycle costs," which include expenses as well as monthly mortgage payments, are lowered considerably in an underground house (see Appendix 3).

This whole idea is summed up by Mark L. Korell, former executive assistant to the Federal Home Loan Board in Washington: "With earth-sheltered homes having the potential to reduce energy costs 30 percent, 50 percent, even 80 percent, they should enjoy special favor as a way to reduce the risk to lender and owner alike of the monthly payment overload."[1]

1. Mark Korell, "Financing Earth Sheltered Housing: Issues and Opportunities," remarks prepared for conference, "Going Under to Stay on Top," Amherst, Mass., June 8 and 9, 1979. For more of this text, see Appendix 2.

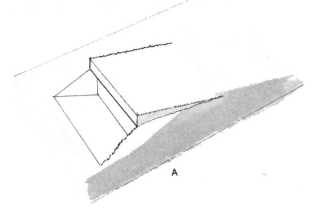

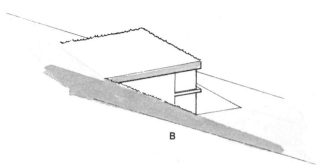

A sloping site might be suitable for a single-level home (A), or a house with more than one level (B). Whenever another story is being considered, some structural problems must be faced.

The belief persists that in-earth dwellings are dark and damp. Loan officers are no exception in many cases. Sketches like this could help.

3d. Homework

Basically lenders like to finance *sound* projects. Your job is to convince them that yours is just that. Educating yourself is the primary task. Do your homework. Next, be prepared to educate your banker. If you *can* arrange a luncheon date, arrive armed with color renderings of your proposed design and charts which forecast how much fuel you'll save. This is the language bankers understand. Don't forget to ask yourself beforehand, Do I really recognize the weak points in this project? You should, because the bank *certainly* will.

Good advice came from a gentleman I met who'd just found financing for a 30,000 square foot underground electronics plant about to be built in Chesterfield, N.H. "Go to the bank and show them lots of pretty pictures, along with charts and graphs that make sense. Get them shaking their heads up and down before you ask for money."

Be a little careful, though. Underground housing might fall under "special interest properties" in the minds of some bankers. This category carries a higher interest rate and a shorter term than a normal residential mortgage. Try to avoid getting into this if you can.

The reputation of the builder you select will have some impact on the lender too. Association with a contractor who has a dubious credit rating will only hamper you. On the positive side, expressing willingness to pay the FHA insurance premium to guarantee the loan should help.

Earthtech 5, Lyme, New Hampshire

3e. INSURANCE

A good sales pitch for building underground might go something like this: by their very nature, earth-sheltered houses are generally more secure from criminals and from wartime bombs. They are less attractive to vandals because they're harder to see. They're safer in a light earthquake, insulated from freezing pipes (see 3f), sealed off from water damage (5g), and protected against weather extremes like tornadoes and hurricanes. This last fact explains why at least twenty-five earthen homes are being, or have already been built in Oklahoma's tornado belt at the moment of this writing.

Theoretically, an earth-covered building, made mostly of concrete, should get the highest fire insurance rating because it represents the lowest risk. Underwriters are slowly recognizing these things, but many cling to the argument that fire insurance costs depend more than anything else on the contents of a building, rather than on its structure.

So the third sad chapter on the newness of earth sheltering reads like this: contrary to wishful thinking, as of yet, there are *not* widespread insurance benefits for underground homeowners. Like bankers and zoning boards, insurance companies are not yet convinced, statistically, that underlivers should be treated preferentially. As the numbers begin to look better, one or more companies will set the trend by offering special premium rates to families living below grade.

This may already be happening. At least two insurance companies, State Farm and Safe Co., are considering 35–40 percent premium reductions for earth-sheltered houses.

3f. MAINTENANCE

An underground house is in, out of the weather. It doesn't need to be repainted or reinsulated. The roof, which is growing grass and shrubs, replaces itself. That means no reshingling,

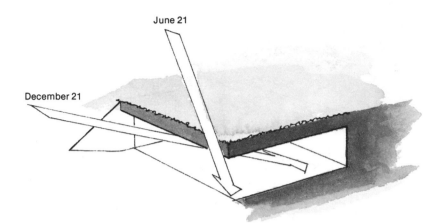

June 21

December 21

Know exactly how far winter sun will reach into the building. This will have an impact on the selection of flooring material. Masonry will release heat during a power failure that halts heating equipment.

although someone may want to mow the roof from time to time.

Because almost everything is buried, most parts of the house are protected from the seesawing expansion and contraction felt by house components above ground every time the weather changes. Expanding and shrinking makes most materials weather and wear out. So do rapid and extreme humidity changes, which don't exist in the soil, but do in the atmosphere.

This fortification against possible freeze-thaw damage, plus special measures to keep the living area bone dry, partly explain why an earth-sheltered house might cost 10 percent more than its above-ground counterpart (see 3g). By the same token, extra waterproofing (5g), careful structural planning (6h), and additional steel reinforcement in the concrete (6h) tend to head off future maintenance costs.

For example: the house below grade will cool off very slowly in the event of a January power failure (9a). On a sweltering August afternoon it will be equally slow to overheat. This is verified by an MIT–University of Minnesota study.[2]

Summertime heat gains in an above-ground house were six to eight times greater than in the underground house to which it was compared.

This relative lack of heat transfer to and from the in-earth structure makes the house not only less dependent on external energy sources, it also lowers the chances of the plumbing freezing during a power outage — a major maintenance expense when and if pipes *do* freeze and burst.

And because the construction either leaves out perishable materials or goes to great lengths to protect them, the house is virtually rot- and termite-proof. The foreseeable life expectancy for a house underground is interminable — without major maintenance costs.

2. Thomas Bligh, Paul Shipp, and George Meixel, "Where to Insulate Earth Protected Buildings and Existing Basements." Paper presented at Amherst "Going Under" conference. For an energy-use comparison between a conventional structure and an earth-sheltered home, see Appendix 1.

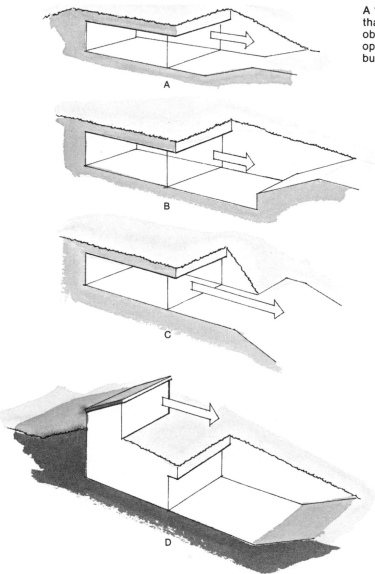

A "bermed" house has earth piled behind it (A). A house that's recessed (B) can have open walls, but the view may be obstructed. A house that's backed into a hillside has an open view (C), while a combination above- and below-grade building might have the best of both worlds (D).

3g. INITIAL COSTS

Nobody, it seems, can put a finger on exact cost figures for underground housing. Estimates vary sharply. John Barnard, of Ecology House, a Massachusetts-based company, claims initial construction savings of 25 percent. The Ecology House, with its distinctive central "atrium" (9d), costs about $27 a square foot to build, including a precast concrete plank roof (6aa) supported by steel beams and columns.

Jeff Sikora, a designer for Housing Research and Building in Swanton, Vermont, just a stone's throw from the Canadian border, maintains that costs to build subterranean homes can be competitive with typical above-ground ranch-style houses. Plans for the "SubT" (9e) call for a well-insulated but non-earth-covered roof structure and a wooden foundation (6v). All but the glazed south wall are buried. His designs estimate out at about $30 a square foot, but Sikora feels he may be able to go as low as $26.

Another Vermonter, Robert Bacon, represents a progressive design firm in Fairlee called Dynamic Innovations Corp. His calculations of $45 a square foot for an underground home are in step with those of Don Metz of Earthtech (1a) and with the

Underground Space Center people in Minnesota (1e), all of whom see below-ground structures as costing an average of 10 percent more. But Bacon envisions costs going down slightly with improved technology and lowered component prices.

Anybody who starts an underground building project with a very tight budget is asking for trouble. Even with extensive site studies (2a), excavation costs, for instance, can easily be underestimated. For a while, professional builders who lack underground experience will contract cautiously, at high prices sure to cover unforeseen expenses. At first, larger contractors who have specialized in excavating, concrete work, and commercial buildings may be less intimidated by underground plans than smaller companies that have restricted themselves to wood-frame residential work.

On the brighter side, if relative price increases for lumber and concrete are any kind of bellwether, trends in building will be more toward concrete. Lumber costs, which are escalating a good deal faster than the inflation rate, continue to make wood construction less and less attractive. Concrete prices are rising more slowly.

In 1980 it's still safe to say that underground houses require more costly materials —

specifically concrete — while above-ground houses are often more labor-intensive, meaning that more man hours are needed to complete the job. But by 1985 underground construction may, in fact, be cheaper.

3h. LIFE-CYCLE COSTS

Let's look at some simple numbers. Statistics are sometimes boring; these are not:

John E. Williams of Hanscomb Associates and Georgia Tech agrees that below-grade building results in a "10 percent construction cost increase." We already know that. But he also claims a "30 percent operation and maintenance cost *decrease*" each year.

In a paper called "Comparative Life-Cycle Costs," delivered at a 1975 National Science Foundation conference on earth-covered buildings held in Fort Worth, Texas, Williams summarized his study of nearly twenty existing buildings in all parts of the United States as well as Sweden. "For each of the six alternative cost configurations analyzed," he said, "the below-grade structure maintained a higher total cost during the early life of the building, and subsequently became a desirable economic

Earthtech 5, south elevation

option later in the life cycle."[3] He added that his findings were "conservative."

The Underground Space Center in Minneapolis (1c) is conservative in its figures, too. It tabulated long-range energy costs for underground houses this way:

Costs to heat the living space in a typical, newly built, 1800 square-foot, above-ground house — call it House A — amount to $440 a year in Minnesota. That's at fuel prices as they were in 1978. (Energy-use figures will decrease in states further to the south.)

This $440 price tag was compared to heating bills in three "typical" underground homes —

call them B-1, B-2, and B-3 — all of which cost 10 percent more to build than House A.

Many claim that underground houses offer energy savings as high as 80 percent for HVAC (the abbreviation for heating, ventilation and air conditioning). House B-1 is not that efficient, however. It uses 40 percent of the energy of House A, at a cost of $176 per year as compared to $440. House B-2 consumes only 20 percent as much HVAC energy as A — $88 a year instead of $440. The third theoretical house, B-3, can be run on only 10 percent as much energy as A. That means $44 a year, rather than $440.

Assuming that fossil fuel costs rise at a rate of 12 percent a year — and that's probably less than they actually *will* rise — and assuming an unrealistically low 6 percent inflation rate,

3. John Williams, "Comparative Life-Cycle Costs," *The Use of Earth Covered Buildings*, p. 55.

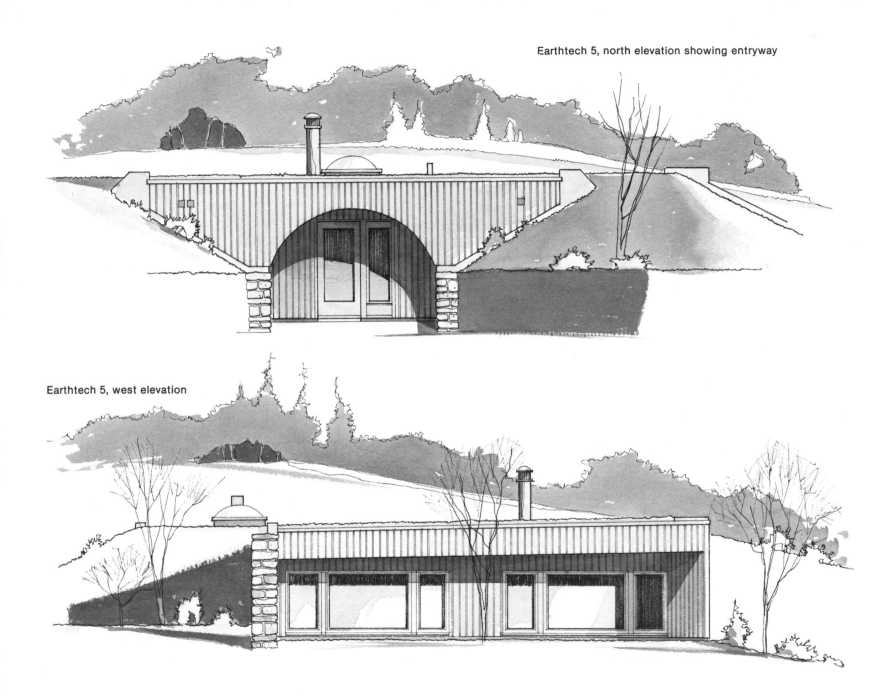

Earthtech 5, north elevation showing entryway

Earthtech 5, west elevation

here's the way home energy expenditures will look after twenty years:

The owners of House A will have spent $3790 for heating.

House B-1 will have cost $1516.

B-2 will cost $758.

B-3 will cost $379.

At this rate, House B-1 will pay for its 10 percent construction cost overrun in energy savings alone after sixteen years. B-2 will break even after twelve years, and B-3 after ten.

Wood heat is a relatively inexpensive — though by no means cheap — alternative to oil, natural gas or propane. It, too, will become more expensive in the next ten years, but its price in many rural areas will probably not soar as rapidly as those of fossil fuels. Passively collected solar energy, which is essentially free once a system's components are installed, cannot inflate in price.

Many underground home designs take these last two factors into consideration. Solar and wood are the primary heat sources (7b). Electric heat is often a back-up system because it's easy to monitor. If the heating units are metered separately from other household appliances, underhomeowners can have an exact record, month-by-month, of how many kilowatt hours of energy they've used. Or saved.

A home-resale analysis for Realtors states that people in this country stay in the same house for an average of seven to twelve years. These figures are likely to change as we move out of energy-greedy dwellings and settle for longer into homes that are cheaper to heat. As the Minnesota study concludes, "The total cost of ownership is considerably less over a period of years with the earth-sheltered alternative." In other words, it'll be worth staying put.

4 EARTH

4a. SOIL

The Feret triangle is a delta-shaped chart of soil types. It has three sides because there are three basic soil classifications used by the United States Department of Agriculture and soil scientists throughout the world. We should all be familiar with it.

Sand, the first type, has visible grains and feels gritty between fingers. When these grains are larger than 2 mm. in diameter, the substance is known as "gravel," a close relative of sand.

Silt is the second major soil type. It, too, has grains, but they're smaller than sand. They can still be felt, but they're too small to see with the naked eye.

The third type, clay, is slippery to touch. A single gram of clay has billions of flat, microscopic particles that stick to each other when dry and slide past each other when they're moistened.

Clay presents problems for gardeners and builders alike. Its composition gives it a plastic quality that allows it to be molded in certain ways (4e). This plasticity explains why clay is known as a "cohesive soil." Unfortunately, a cohesive soil is also one that expands when wet. Expansion of this kind plugs up most drainage passages in clay. It also exerts enough pressure against a weak concrete wall to crumble it like dry breadcrust (6n). Clay is *not* the soil to have against an underground wall.

Most soils are combinations of these three types. "Loams," in fact, are mixtures, although a "sandy loam" is very different from, say, "silty clay loam." Combinations are almost infinite, depending on the chemical make up of the

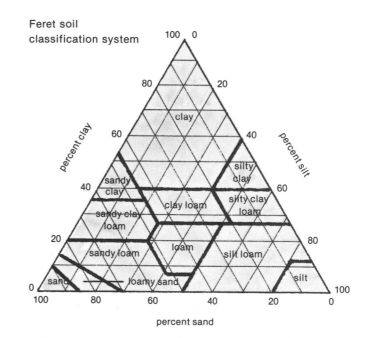

Feret soil classification system

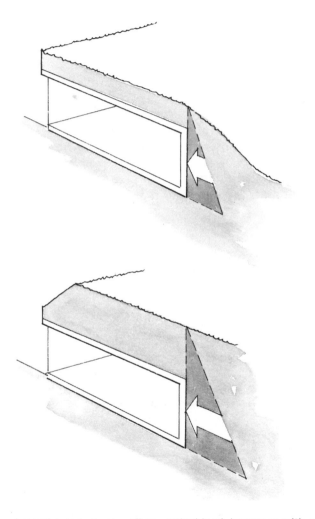

Lateral pressure, as well as vertical load, increases with soil depth.

ingredients — making soil science a complex subject. Many of the various soil blends are catalogued and named, and the list seems endless.

The wise underground builder takes a hard look at soil as a site is being planned (2d). If the character of the earth presents major problems that can't be solved by proper excavating (4f), and backfilling (5f), consulting a soil engineer may be necessary.

4b. Bearing Strength

Andy Davis, an electrician, thought about its design for a long time before he started building Davis Cave (9a). He knew it would have to be strong to withstand earth pressures. Among his other computations, he found that soil would need a weight-bearing capacity of 3000 pounds per square foot to support footings for his earth-sheltered house. Davis Cave should never settle, thanks to its owner's good planning.

A look at *Architectural Graphics Standards* by Ramsey and Sleeper, a reference text for designers and builders, suggests that Davis needn't have worried quite so much about bearing strength. Bedrock they say, has an allowable bearing of 200,000 pounds per square foot. That's 100 tons. "Foliated rock," such as bedded limestone, or hard granite "schist," can support forty tons/ft². Gravel and hardpan can hold up to ten tons/ft², loose gravel and coarse sand four tons/ft², and even wet sand, like that on a beach, can bear two tons/ft². Most underground homes will not exert more than the 1.5 tons (3000 lbs) estimated by Davis.

The only soil to offer unsuitable bearing would be soft clay, with a maximum capacity of one ton/ft². Builders' fill, highly cultivated topsoil, or organic soils like swamp muck are incapable of supporting almost any kind of building.

Upper soil layers should be excavated deep enough to establish a flat "pad" in "undisturbed subsoil" — the condition specified for most conventional construction. In short, almost any soil should be acceptable support for footings and grade beams, particularly if it's mechanically compacted first.

4c. Thermal Conductivity

Tests at the University of Minnesota measured heat transfer through five distinct soil types. The conclusion: soil type has little influence on its ability to conduct heat. The important factor is moisture content.

Thermal conductivity — the soil's tendency to transport heat from one place to the next — can

SOIL CLASSIFICATION BY PLASTICITY INDEX

TYPE OF SOIL COHESIVENESS		DEGREE OF PLASTICITY	PLASTICITY INDEX	LIQUID LIMIT	PLASTIC LIMIT
sand	non-cohesive	non-plastic	0	20	20
silt	partly-cohesive	low-plastic	<7	25	20
silty clay	cohesive	med-plastic	>7	40	25
clayey silt	cohesive	med-plastic	<17		
clay	cohesive	high-plastic	>17	70	40

Source: Alfreds R. Jumikis, *Introduction to Soil Mechanics* (New York: Van Nostrand, 1967). Reprinted with permission.

change as much as tenfold with changes in moisture content. The wetter it gets, the better it conducts. The importance of knowing the moisture-holding capacity of a site's soil has already been stressed. More important is seeing that soil adjacent to concrete walls be porous enough to drain well. Sand and gravel are best. Stated most simply, the better the drainage system (5f), the less heat will be drawn away from the house.

Fine-grained soils close to the surface are more susceptible to frost heaving because they hold water more easily. Ice is an even better heat conductor than damp soil. Wet topsoil banked against a concrete wall, even above porous backfill, can become a major heat-loss artery.

Soil temperature also varies according to a number of other factors, namely, latitude, site characteristics, weather conditions, and the nature of soil particles themselves. But below eighteen to twenty inches, soil shows very little temperature fluctuation from day to day, regardless of what the temperature on the surface is doing. Deeper still, soil temperature rises and falls even more slowly — only with dramatic seasonal changes.

Each foot of soil means a seven-day time lag in temperature variation. At eight feet it could be two to three months before soil particles show a transition from summer warmth to wintertime cool. Below twelve feet temperature extremes hardly exist at all.

Soil depth is a great heat stabilizer. Normal soil temperatures well below the surface of most parts of the globe average about 55° F. year round.

All of this reduces the number of heating degree day units needed to raise the temperature in underground living space (see chapter 7). Earth-sheltered homes are best suited, then, to areas where climate varies from season to season — places where people spend lots of money on insulation and energy, either to warm or cool themselves.

4d. Pressures

It's not easy, sometimes, to think of soil as a fluid. It is — though it's far more dense than most "liquids" we see every day. Soil flows with water, settles with gravity, shifts in changing winds. It's moved and smoothed by ice — by glaciers bigger than continents or by one tiny crystal shard.

Soil's density is changed by heat, by moisture, by the countless organisms that live in it, and by its location and depth. It's deflected off things

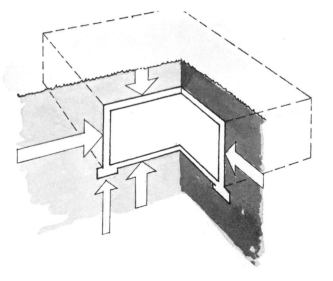

Pressures against an underground house come from all directions.

harder than it — by denser soil, bedrock, and strong man-made buttresses (see 6h).

We live on top of a vast, moving ocean of earth. The movement is imperceptible most of the time, but it's there. As we plunge deeper into the safety of the soil, we have to recognize the immense pressures that exist there (6i).

Tension among individual particles of soil may be so great in places, that we can dig under it, letting it arch overhead and support its own weight (6m). Elsewhere the introduction of moisture will destroy this bond between soil particles, creating a collapsible soil that can support little weight. Chemicals can build and relieve soil pressures in cunning ways. Salts in seaside sands, for instance, can slowly corrode a concrete wall, aiding the soil's force, weakening the wall's.

Lateral wall pressures must be a large consideration in underground design — much

more so than above the surface. In parts of Texas, Colorado, and throughout the western United States, expansive clay, which swells when wet, is common. This soil has been known to exert more than 30,000 pounds of pressure against a single square foot of vertical wall. And it can effortlessly heave and buckle a concrete slab from beneath.

In a sense, our in-earth homes will be immobile submarines. We can't venture into dangerous and uncharted soils. We have to consider depth, and barricade ourselves against the stresses and currents in the ground around us.

4e. Slope Stability

We'll often make waves when we earth-shelter a house. A mound of earth-fluid — called a *berm* — designed to protect us from cold, faster moving air-fluid — called *wind* — can only be built within certain limitations.

How a soil naturally rests in place is its "angle of repose." The angle of repose — its maximum angle of stability — depends on how the soil particles rest against and on top of each other. Obviously we don't want to build berms that are too steep to be stable.

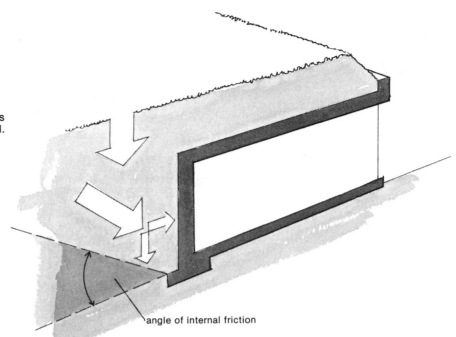

Earth loads deliver wedge-like thrusts against a wall.

angle of internal friction

Clay, with particles that stick together, can have a steep angle of repose so long as it's dry. But water-fluid can coax it to flatness. Twenty-six degrees, then, is about the steepest pitch for most soils, and granular soils will probably assume an even shallower slope. If earth needs to be piled up steeper than this, a retaining wall will be needed to keep it there (6v).

Because overly loose soils present inherent slope stability problems, they're often difficult to excavate (4f). The ground keeps flowing down around the digger — man or machine —

sometimes faster than the digger can dig. Be familiar with the at-rest characteristics of the soil on your site.

4f. EXCAVATION

Good excavation is a paradox. Sane earth moving calls for brute force, but it removes only as much soil as necessary, disturbing the surroundings and those living things that grow out of the earth as little as possible.

In the best cases, the right machine, matched with the right surgeon-like operator, is matched to the right job. Nearby bark doesn't get pierced, shrubs don't get trampled, and most grass is left unstripped. The operation is a

TYPICAL ANGLES OF REPOSE

SOIL TYPE	CONDITION	IN DEGREES	AS RATIO
Clay	dry	30°	1.7:1
	damp, plastic	18°	3:1
	wet	16°	3.5:1
Earth	"vegetable soil"	28°	1.9:1
	loose, or humus	30°	1.7:1
Gravel	average	32°	1.6:1
Sand	clean	33°	1.5:1

Excerpted with permission from K. Labs, "The Architectural Use of Underground Space," unpublished thesis, 1975. The thesis is available from Kenneth Labs, Star Route, Mechanicsville, PA 18934.

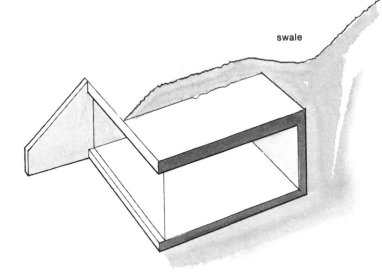

swale

Surface water can be directed around the house by contouring, or "swaling" the earth shelter.

success only when the house has been implanted, the earth's wound healed, and greenery has covered the scar.

In the end, we should never know that chain saws, bulldozers, backhoes, and concrete trucks were there.

Any trees that are cut should be moved far enough away from the house site to be out of the way. Firewood should be saved. Branches and brush might be chipped into mulch for later use (4k), or judiciously buried someplace apart from the foundation hole. Large stumps may have to be extracted. Rich topsoil shouldn't be mixed with infertile subsoil, but should be stockpiled separately.

Veins of water exposed by digging can be rerouted into the earth through drainage lines made of tile or plastic (5f). Underground water can also be diverted to some place where it's comfortably accepted, and will bring no discomfort to those in the house. Dirt piles should be tactfully located, to allow clumsy concrete trucks to snuggle in close to wall forms at pouring time.

Excavation in ledge can be more difficult, but it's neither impossible nor prohibitively expensive. Two Metz-built houses, Baldtop Dugout and Earthtech 5 (9f), are set partially in bedrock. In both cases blasting costs never exceeded $1000.

Using dynamite *will* mean bringing in drilling equipment and a large compressor. Holes 1½ inches wide will be drilled six feet or more into the rock, and about six feet apart in all directions. These shafts will be packed with dynamite and sand. The holes will be laid out in either a "V" pattern or a grid. Each is wired with a small blasting cap near its top.

If there are other buildings nearby, a heavy steel mat may be laid over the dynamite — to muffle the noise and keep stray rock fragments from flying off in any direction. But airborne missiles like this are not usually expected.

The blast should be like a stone dropped into a mill pond, radiating from the center of the ledge in circular waves. Blasting caps near the

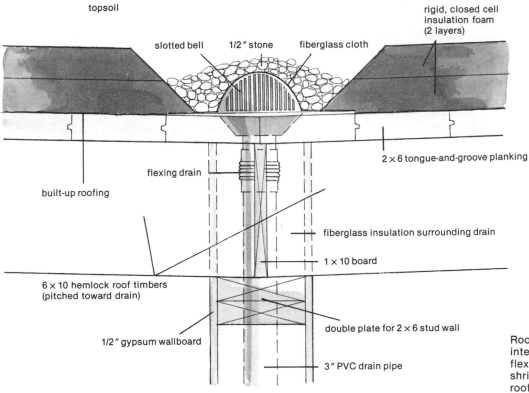

topsoil

rigid, closed cell insulation foam (2 layers)

slotted bell   1/2″ stone   fiberglass cloth

2 × 6 tongue-and-groove planking

flexing drain

built-up roofing

fiberglass insulation surrounding drain

1 × 10 board

6 × 10 hemlock roof timbers (pitched toward drain)

double plate for 2 × 6 stud wall

1/2″ gypsum wallboard

3″ PVC drain pipe

Roof water may be drained within interior wall partitions, through flexing drains that permit some shrinking and twisting in the heavy roof timbers.

middle will go off first. Those further to the outside are set at later time intervals, triggering explosions microseconds after those in the center.

When the dust settles — if all goes well — the ground in the excavation will look like a just-risen loaf of bread. This new mound of fractured rock can be removed, crushed further to be used for backfill, or trucked away. Once the base of the hole is made flat, and found to be about three feet larger than the building, the house footings can be laid out.

4g. ROOF SOIL

Soil on a roof, like soil next to a wall, must drain well. This is usually accomplished by spreading topsoil over a layer of sand or gravel. The sand sits atop foam building insulation (6y) which, in turn, covers a waterproof membrane.

The roof of Earthtech 5 looks essentially flat (9f). Actually, Don Metz designed it to pitch slightly — about a quarter-inch per foot. Here

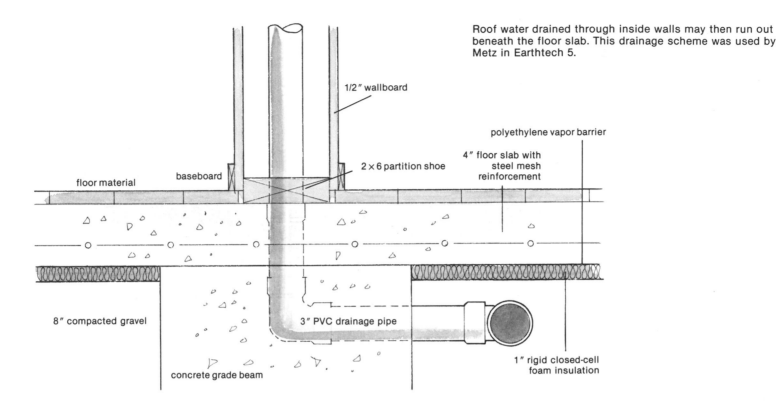

Roof water drained through inside walls may then run out beneath the floor slab. This drainage scheme was used by Metz in Earthtech 5.

1/2" wallboard

polyethylene vapor barrier

4" floor slab with steel mesh reinforcement

2 × 6 partition shoe

floor material

baseboard

8" compacted gravel

3" PVC drainage pipe

1" rigid closed-cell foam insulation

concrete grade beam

water flows to drains running vertically through an interior wall of the house, then out beneath the floor slab.

Any soil gets heavier when it's wet, of course. And it's fully expected that the heavy wooden roof structure of Earthtech 5 will settle and give as the weight of soil — or snow, or vegetation, or people — on the roof changes. The tops of the in-wall drains, then, pass through Flex-i-Drains, manufactured by the Johns Manville Corporation.

The Flex-i-Drain permits the 6" × 10"

hemlock timbers, which may be installed when they're still "green," to shrink and twist as they dry out and settle into place — all without risk of pushing the drain flange up through the roofing membrane. The top of the Flex-i-Drain itself — above the roof — is a slotted bell to be covered by a fiberglass mesh filter.

There's no reason to fear plant roots will destroy a roof membrane or clog drains. Roots are in search of moisture and nutrients, after all, and neither of these exists below the waterproofing layer or in a vertical downspout.

What's planted on a roof depends largely on the depth of soil there. Topsoil should drain readily above a layer of sand or gravel, and it should probably be mulched at first. More than five feet of soil would be needed to support large shrubs or small trees.

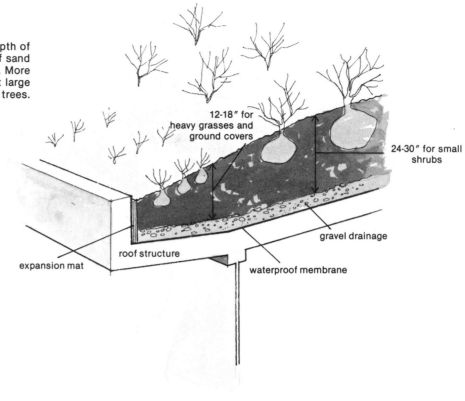

12-18″ for heavy grasses and ground covers

24-30″ for small shrubs

gravel drainage

expansion mat

roof structure

waterproof membrane

4h. Depth

Common question: how much soil should there be on the roof? Standard answer: one to three feet seems best. More or less than this, and the trade-offs become poor. More soil gets extremely heavy, stressing the roof excessively. Lightening agents such as perlite, vermiculite or styrofoam beads may be added to the soil, but their benefit is questionable. Less than a foot of soil provides too little thermal mass (7a) to make the earthen roof worth the effort and limits the growth of vegetation.

Rooftop soil depth will have an impact both on how the house retains its heat and on what can grow there. This rough formula is offered by Thomas Wirth, a consultant to the Underground Space Center: twelve to eighteen inches, he says, will support heavy grasses and low ground covers. Twenty-four to thirty inches are needed for small shrubs. Large shrubs and small trees will need a soil depth of more than five feet to survive.

A local nurseryman or landscape architect would be an excellent source of advice about the best grasses and shrub varieties for rooftops in your area.

4i. Vegetation

Native plants, for the most part, will do better than trees, bushes, or low ground covers imported from even a slightly different climate. Rooftop vegetation may be started by borrowing plants from another part of the building site. To assure fertility, the topsoil might be laid over a layer of well-rotted manure, which will also drain.

Vegetables can grow on a roof as easily as lawn, though they'll need more care. Edible plants make a lot of sense, particularly since

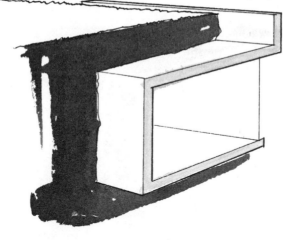

concrete parapet

Wells reminds us that a large parapet designed to retain earth on a roof is the "wrong approach." He goes on to say, "That's only asking for trouble. Expansion and contraction, plus ice damage, can take all the fun out of a building like this. And the massive structure can bleed a lot of heat, too."

extended by warmth from the living space below it. Would it be ready to plant earlier in the spring? Could autumn frosts be postponed a few days? Many believe so.

4j. ROOFSCAPING

The temperature of roof soil will be affected by the nature of its cover vegetation. Any intelligent dog knows that a patch of grass is cooler than a blacktop parking lot on a hot day. What the animal does not understand, probably, is why. Green growing things shade the moist ground — cooling in itself. They also transpire — gradually giving off some of their moisture to the surrounding air, thus cooling themselves. Transpiration is to plants what perspiration is to people.

In some underground homes, the overhanging roof — meant to protect southside windows from too much sun — takes the form of a trellis. Vines, growing down from the trellis, shade windows and exposed walls even more. Ingelman ivy, which sports large leaves in summer, sheds its foliage in the fall, just in time to let solar radiation through to the windows again. It's often recommended.

that great American tradition known as The Lawn has come under fire recently from environmentalists. Lawns, they argue, are "sick and counterproductive energy sinks." Ken Labs, architect, quotes Dr. Frank Engler, ecologist,

> Suburban landscape is already the domain of lawns, lawnmowers, subsoil called topsoil, alien species that are susceptible to every known form of pest, disease, and inherent weakness. The whole is kept alive by an inordinate amount of expensive care and attention, comparable to our practices in homes for the aged and incurable.[1]

Apparently roof soil will hold a certain amount of heat generated within the house. Someone should investigate if, in fact, and how much, a rooftop gardening season might be

1. Kenneth B. Labs, *The Architectural Use of Underground Space*, p. 116.

The central atrium within John Barnard's Ecology House in Osterville, Massachusetts, is protected by a fence. By now this artificial barrier is concealed in flowers and shrubbery.

4k. Erosion Control

Roof soil might also benefit from a buffer zone of organic mulch, especially before a thick blanket of vegetation has time to establish itself. Mulch, whether it be wood chips, grass clippings, chopped hay, or any of the many other choices, will slow rooftop erosion and help the soil retain its wetness. The mulch will gradually decompose, and will need to be replenished every so often, until there's healthy plant growth.

At first the Wells underground office in New Jersey had "rottable roof curbs" for erosion control. These were made from lumber scraps left over from the construction of the building. By the time they rotted away, the vegetation was strong enough to keep the roof soil from washing away.

4l. Barriers

The better the roof planting, the harder it is to see the roof. And as Malcolm Wells (1h) is fond of saying, "When roofs reach the ground, kids reach the roof."

Obviously it's best not to have people — kids or adults — walking off the edge of a roof or falling into the central courtyard of an underground house (9d). Fences to prevent such mishaps can be disguised and camouflaged in shrubbery. Better still, a close-order formation of thorny bushes creates its own natural barrier.

5 WATER

5a. SURFACE WATER

Dealing with water we can see is one thing.
Controlling water that's underground and out
of sight is something else (5b). Surface water
can usually be directed off and around a
subterranean house without major difficulty.
The simple principles are the same as in
overbuilding — pitch and gravity flow.

Slope earth away from the
house wherever possible.
When it's not, and low spots
exist, drainage lines will be
needed.

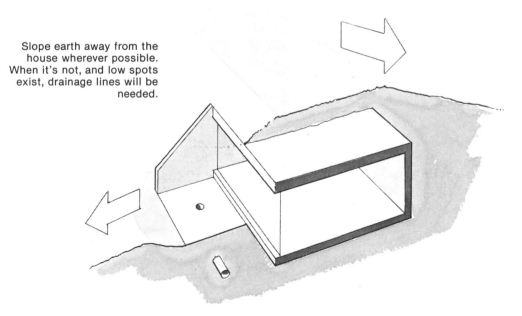

Backfilling, landscaping, and berming (4e) must
slope away from the house in all directions.

One to five degrees is adequate pitch for
removing surface water by gravity. If a soil
drains well, and sufficient man-made drainage
surrounds the walls, the ground surface near the
house can slope more steeply.

But how will diverting, slowing, or
accelerating the water patterns affect a downhill
neighbor? Will you flood him out by suddenly
sending him more water than he can accept? Or,
will you deprive him of water he counts on?

"Ponding" describes a condition where water
has no place to go. And a roof is obviously not
a good place for standing water. There are two
ways to eliminate surface water from a rooftop.
The first is to pitch the roof — and/or the soil
on it — so water runs off to the side or drains
away through partitions inside the house.

The method of directing roof water down
within interior walls has an advantage in that
the roof may slope away from its overhang.
This way there's no drip line in front of
windows on the exposed side of the building.
Flex-i-Drains solve problems of shrinking roof
timbers or precast concrete roof planks, which
might otherwise endanger in-wall drain pipes.

A second way to reduce rooftop moisture is
to utilize it by allowing plenty of vegetation to

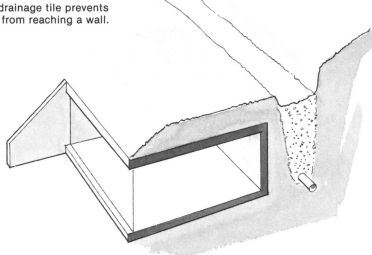

A swale and a gravel trench with drainage tile prevents hillside runoff from reaching a wall.

grow there (4i). If plants are healthy and have heavy root systems, they'll intercept much of the water before it ever reaches the roof's waterproofing membrane (5i).

5b. UNDERGROUND WATER

Once water disappears below the surface, it's harder to trace and to direct. Moisture movement through soils is complicated and not completely understood. But it's clear that water is driven away from heat sources in the earth, toward cooler places. Here we find another benefit to living underground.

Heat travels in soil as it travels elsewhere — moving from warm space to cold space. The question is not *whether* heat will move, but how fast this transfer will take place. The wetter the soil, remember, the better it transfers heat (4c).

In winter, after surface soil freezes and no longer allows water to sink to lower layers, earth next to a heated wall will be warmed. As it dries there, the ground is a less effective heat conductor. So more warmth stays in the house.

In summer, this drying effect isn't present because the walls contain less heat. Some moisture can drain into the ground near the house. Now, heat *is* transferred through this

moisture to the cooler soil nearby, lowering the temperature inside the walls. Natural air conditioning.

5c. Water Table

The "water table" describes the maximum ratio of water content to soil volume. It's defined as the highest level at which soil is saturated with moisture. It doesn't stay at the same depth, but moves up and down, somewhat, according to recent amounts of precipitation and the ground's ability to drain. When the water table rises above the top of the soil, the earth there is said to be "drowned."

The average level of the water table will be a determining factor in deciding whether an underground house is feasible on a given site. A

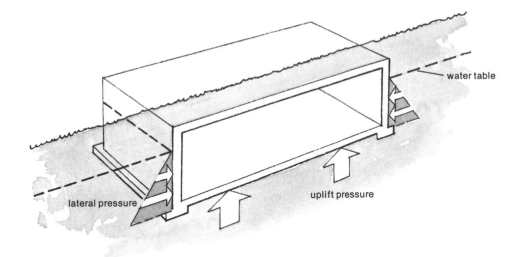

water table

lateral pressure

uplift pressure

A high water table not only exerts increased pressure against walls from the side, it also creates uplift pressure against the floor slab from below.

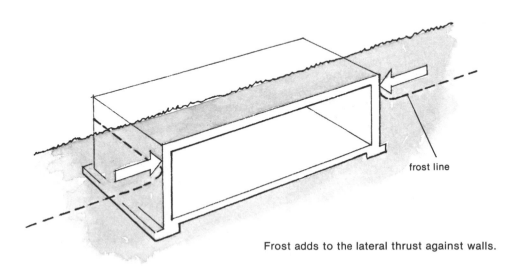

frost line

Frost adds to the lateral thrust against walls.

high water table doesn't necessarily eliminate a plan for building below grade, but it usually spells more dollars.

In some instances a water table can be lowered by adding more artificial drainage (5f). More often than not, however, those who build in soaked soil struggle constantly with water that's trying to push down walls and "float" floor slabs by pushing up from below. Special — and expensive — measures must be taken to counteract these pressures.

5d. Frost

Soil moisture problems are compounded by freezing temperatures. Soil expands with the addition of water, and it expands still more when this water turns to ice. Next to expansive clays, silty soils are the worst in this respect, expanding the most.

The tops of footings and grade beams must be established below regional frost line depths. While a footing might be placed directly below a floor slab in Florida, for instance, it may need to be six feet deeper in Alaska.

Most severe wall pressures may occur in the spring. Soil nearest a wall will thaw first, with the help of heat from the house. Still-frozen earth further away will be unable to soak up water. Runoff from melting March snows or April showers has no choice but to settle down next to a wall, taxing the drainage system more, possibly, than at any other time. If the drains fail, the area around the footings becomes a catch basin for water.

5e. Hydrostatic Pressure

When water accumulates to the point of saturating a soil, the water table rises. Below this saturation line, ground water will place 62.4 pounds of pressure per square foot against a structure for every foot of depth. This water force, which is exerted in all directions, is known as hydrostatic pressure. At eight feet below the water table, hydrostatic pressure alone amounts to 499.2 lbs/ft².

To resist potential pressures like these, walls and floor slabs need reinforcement. Earthtech 5, for example, a typically sized earth-sheltered house, has eight-inch walls with ½-inch steel bars running vertically and horizontally, sixteen inches apart.

The floor slab, four inches thick, is poured over eight inches of thoroughly compacted gravel with a sand topping. The concrete rests on a one-inch layer of foam insulation and a sheet of six-mil polyethylene which serves as a

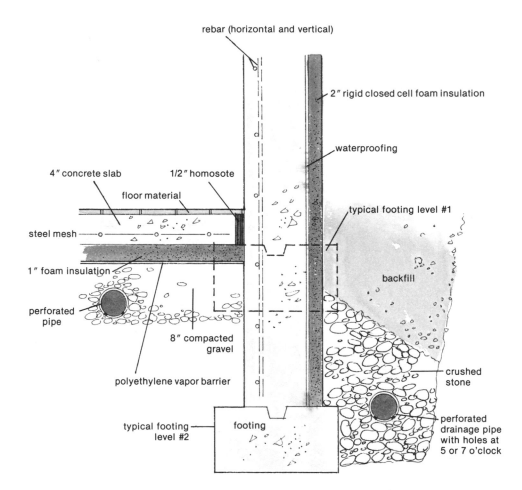

Typical wall base and footing cross-section (after Metz)

vapor barrier (5g). A 6″×6″ steel wire mesh, set within the slab, is meant to eliminate shrinkage cracks (6n).

None of the five Earthtech homes has been built on a site with a dangerously high water table. Had one been, lines of drainage tile might have been placed in the compacted gravel below the floor slab to encourage water movement away from the house. More steel reinforcement might have been added as well.

5f. DRAINAGE

Proper drainage, as we already know, is vital to an underground house. Tiling, the standard method of drainage that surrounds foundations of conventional houses, can also be effective in below-grade homes. Regular four-inch perforated drainage pipe, sometimes called "tile," should be placed in a bed of 1- to 1½-inch crushed stone at the base of the concrete footings. The holes in the pipe face toward the five or seven o'clock position, not directly down.

Drainage tile, which might be PVC plastic, runs around the perimeter of the house, and must slope gradually downhill to permit gravity flow. Couplings can be covered with gravel and

tar paper to prevent fine sand and silt from clogging the line. As ground water is moved away through these footing drains, it must be absorbed in a dry well or leach field. On hillside lots, these lines can "daylight" to the surface naturally, somewhere below the house. Freezing at these ends could be a problem in cold climates, however.

Backfill above the tile must be free-draining gravel or coarse sand that filters out any fine soil particles and silt that could plug up holes in the tile. This backfill might be compacted every two or three feet or so, then capped with regular, less-porous topsoil that pitches gently away from the wall.

Engineers often advise that wherever possible, ground water should be drained by gravity rather than mechanical pumps. It's less expensive and more reliable. Gravity never breaks down.

5g. WATER AND VAPORPROOFING

Concrete and masonry, reinforced or not, shrinks and cracks as it cures. None is waterproof. Whatever sealant is used to keep water out of an underground house must be elastic or it also will crack and leak. Pressure-treated lumber, not as strong as concrete and steel, is frequently used to build underground walls (see 6s). It, too, needs to be waterproofed.

Almost invariably, soil that encases an earth-sheltered house contains more moisture than does the air inside the building. Capillary action entices this moisture toward and through the walls, raising the humidity level indoors. Because too much humidity is uncomfortable and harmful, the walls must be vaporproofed as well. Unfortunately, not all "waterproof" membranes are also vaporproof.

Some dampproofing products are designed to break this capillary action (5h). And a system developed in Sweden uses a regular waterproof membrane in combination with "rock wool," a mineral fiber insulation, to discourage moisture from approaching a wall to begin with.

5h. Sealants

When a wall is "parged," a coat of dense stucco-like cement is troweled or brushed over the outer concrete surface. But before *any* sort of waterproof coating is put on a wall, all small cracks and snap-tie holes, left over from concrete forms, must be filled with material that's compatible with the waterproofing.

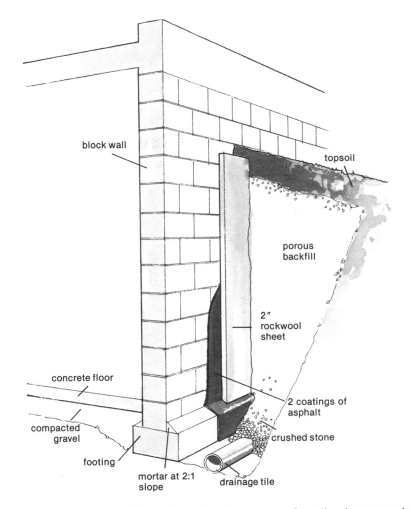

block wall

topsoil

porous backfill

2" rockwool sheet

concrete floor

2 coatings of asphalt

compacted gravel

crushed stone

footing

mortar at 2:1 slope

drainage tile

Theoretically, a waterproof membrane is not necessary. Scandinavian researchers have found that air spaces in a 2-inch rockwool sheet, for example, will interrupt capillary draw provided there's not more water than can be drained from outside the wall. If these air spaces become saturated, though, this Swedish system is not effective. The double asphalt coating to a height of 20 inches above the footing allows for a small build-up of water pressure.

Thoroseal is just one well-known brand name for a waterproof parging designed to both seal and break capillary action near a wall. Its manufacturer recommends that it be applied on the outside of the wall in one or more coatings.

Like concrete, rigid parges will contract in time, cracking in the process. Once they do, they're no longer waterproof *or* vaporproof. Needless to say, patching cracks is impractical

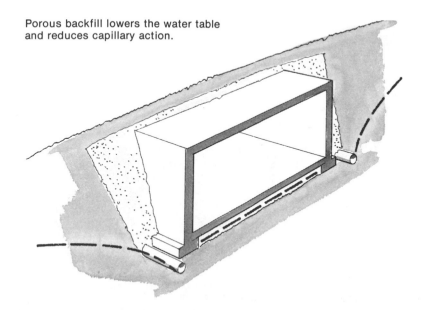

Porous backfill lowers the water table and reduces capillary action.

once an underground house is backfilled and bermed.

Asphalt has been used for many years as a "waterproof" foundation coating. It, too, can be brushed on or troweled. Also sprayed. More recently the quality of asphalt has deteriorated as petroleum companies extract more and more chemicals from it to be used for other purposes.

Asphalt, actually, is water-soluble over the long haul. It may eventually break down, returning gradually to the soil emulsion nearby. When a foundation coating on an above-ground home disintegrates, the basement is damp, but the residents stay high and dry. Below ground, when asphalt alone is used as waterproofing, people are low and wet.

The Andy Davises (9a) used a bituminous pitch sealer painted on the concrete walls of their cave dwelling. On Ecology Houses (9d), John Barnard has used sixty-pound asbestos and hot pitch on the roof, and hot pitch alone on the concrete sidewalls.

Pitch, however, presents some problems. In the first place, it's considered potentially toxic by the Occupational Safety and Health Administration (OSHA). In the second, pitch, like asphalt, tends to become brittle after a time, and will fracture with the concrete it's intended to coat.

Most commonly used sealants aren't flexible enough to bridge and reseal cracks that can develop after they've been applied. Happily, there are better waterproofing alternatives.

5i. Membranes

A polyethylene membrane is often used as a vapor barrier in above-ground studwall construction. Clear "poly" sheeting is also excellent beneath a poured floor slab (5e). But it shouldn't be considered a satisfactory waterproofing for concrete walls. When it's exposed to ultraviolet sunlight its entire composition is threatened: soon whole sheets are turned to fragile shreds.

Black polyethylene is better in this respect because ultraviolet rays threaten it less. But it too can be easily punctured by stones during backfilling. All polyethylene is difficult to seal where it overlaps, and can't resist much water pressure where sheets are joined. Even when embedded in sticky mastic, and *sealed* with mastic at its joints, polyethylene is never quite dependable.

A "built-up" membrane consists of alternating layers of tar paper and tar — usually asphalt and felt roofing paper. When it's done properly on a wooden roof structure that drains

readily, it's better than satisfactory. In fact, built-up roofing has been specified on Don Metz's early Earthtech houses (9f). This membrane is often used on flat roofs in combination with flashing and plastic cement around skylights (8e), chimneys and stacks that pass through the roof. Then again, it usually can't be trusted on a vertical concrete wall. Because the asphalt comes hot out of the kettle,

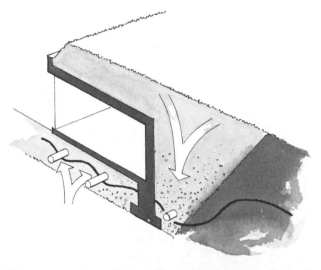

Water approaches a house from above and below. Drain tile has a profound effect on the "downdraw curve."

The Alcorn House, by Don Metz, was still under construction in August of 1979. The greenhouse entryway connecting the above-grade garage and the enclosed atrium had not yet been glazed. But the bituthane membrane had been installed on the roof before four inches of Styrofoam insulation had been laid down. Two chimneys, two skylights, and several Flex-i-Drains, visible here, passed through the roof. Sand was added next, then the final layer of topsoil.

and must be mopped onto the wall, it's difficult to spread with perfect evenness.

Bituthane — polyethylene-coated, rubberized asphalt — sounds impressive, and it is. It's relatively inexpensive, and studies have shown it will elongate as much as 300 percent even at temperatures as low as –15° F. In other words, stretching it can't break its seal. Applied at about $1.00/ft^2, bituthane is an excellent waterproofing for wooden roofs.

Butyl rubber sheeting seems to be the best synthetic waterproofing membrane available. It stretches across cracks without weakening or breaking. Butyl membrane is $\frac{1}{16}$ inch thick, and can be delivered in large rolls to reduce the need for seams, which are still difficult to seal and make entirely watertight. It's also expensive — $1.50 to $2 a square foot at 1980 prices.

It's already been suggested that a waterproof membrane such as butyl rubber can be damaged when the walls are backfilled. Later it can be attacked by frost, acidic soil, and possibly roots.

For protection it can be coated on the outside with asphalt, or covered by noncompressive insulation (7e), felt paper, fiberglass, or asbestos cement board (9f). When it's applied inside insulation — on the "warm" side — a waterproofing membrane does double duty as a vapor barrier.

5j. Bentonite

We know about expansive clays and the pressures they can place on walls (4d). One of life's small but profound lessons is that "bad" things can sometimes be "good." Bentonite is a natural material that can never decompose. It's an expansive clay, and it may be the best water-vaporproofing of all. It's used to seal pond bottoms in gravelly rural areas. Here it withstands tons of hydrostatic pressure. It spans cracks and reseals leaks long after it becomes a pond sealant *or* a wall coating. The more it's exposed to water, the better it waterproofs.

Bentonite can be ordered in easy-to-apply cardboard panels, or it can be sprayed on walls in layers 3/8 inch thick. The spraying might be done by a licensed "Bentonize System" applicator, and this commercial process carries a five-year guarantee. Bentonite can also be applied with a mason's trowel.

Like other clays, Bentonite can be washed away in a heavy downpour. Once it's applied, the wall should be backfilled fairly soon.

6 DESIGN

6a. CONSIDERATIONS

Everyone agrees: plan, plan, and plan. Then review the plans some more. Because earth-sheltered design is still in its adolescence, because underground space must be used wisely, because changes and additions will be expensive and difficult later on, and because the house should fit the contours of the site instead of creating its own shape, a sensitive underground architect must be a key member of any planning-building team.

At one time or another, all designers need to be reminded of a timeless architectural maxim: a truly economical structure is one in which the components serve more than one purpose. A partition, for example, is used primarily to divide space. But it can also bear some of the roof load, and serve as an enclosed passageway for water draining off a roof (4g).

Along the same line, a concrete and tile floor is more than a flat base for living space. It's a shield against the forces of underground water (5e). In addition, it can be a solar collector and solar storage area. Exterior walls keep outside pressures outside, hold up roofs, and play supporting roles as heat banks. And a roof can be a garden as well as shelter.

Unless there's money to burn, extravagant frills like the indoor-outdoor pool in Frank Lloyd Wright's Solar Hemicycle (1g) should not even be considered. Open water sitting behind glass can produce monumental evaporation and humidity problems.

At the opposite extreme, a plumbingless composting toilet, such as the Clivus Multrum, may seem like an ecologically sound idea. But it has inbred problems, too. Unless a two-level house is planned (in spite of the structural drawbacks presented by a second story), there will be no access to the toilet's composted waste without further excavation below the floor slab.

In short, to keep costs within reason, simple form must follow simple function in all parts of the house.

6b. BASIC ARCHITECTURAL TYPES

There are three fundamental design possibilities for an earth-sheltered house:

The "elevational" concept involves a semi-recessed building backed into a sloping site. Windows appear only on the exposed elevations — preferably on the south side; possibly on the east or west. The north and east sides are covered with earth, as is the roof. An elevational structure is instantly compatible with passive solar heating (7f). It's a ready-made solar collector, in fact. Floor plans for an elevational house must be done carefully if each room in the house is to have a view of the outside world through a vertical window. Spaces deep within the house can be lighted through skylights or lightwells (8d).

A flat site might lend itself to a fully recessed house, built around a central opening or courtyard. This is an "atrium" design, modeled after the traditional home of ancient Rome, which had a central room, open to the sky, with porticoes opening to all sides. Today the length, width, and depth of the atrium has a profound effect on lighting, ventilation, and solar heating possibilities in an underground building. All habitable rooms can receive natural light from the central courtyard (9d).

In a "penetrational" design, natural light and fresh air reach the living space through gaps either left open or dug into the earth around the perimeter of the structure. The penetrational concept offers a second answer to the problem of a nonsloping site. But these penetrations reduce the effectiveness of earth covering. A home of this design will also need more retaining walls (6y) than either of the other two types.

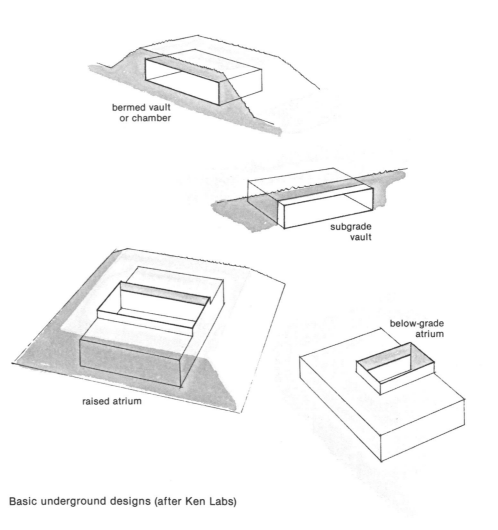

bermed vault
or chamber

subgrade
vault

raised atrium

below-grade
atrium

Basic underground designs (after Ken Labs)

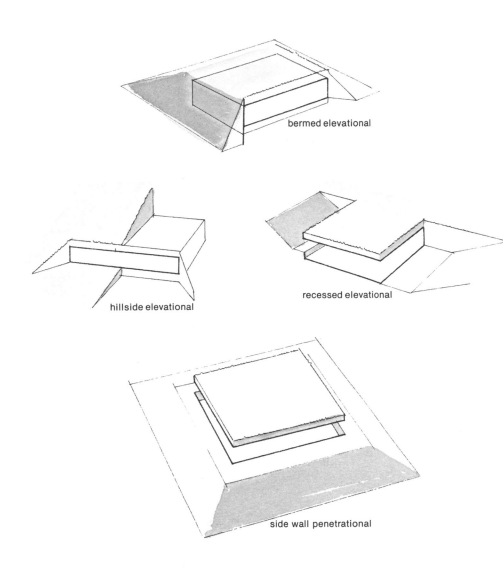

bermed elevational

hillside elevational

recessed elevational

side wall penetrational

Stu Campbell

The living room door opens into the courtyard of John Barnard's first atrium-style Ecology House.

6c. Surface Area

Although it's still important, external surface area is not nearly as big a concern in plans for underground houses as it is in above-ground shelters. A peek into a crystal ball suggests that as terratecture is refined, earth-sheltered structures will enclose larger and more rambling spaces. These inner spaces will be protected by more glass, representing more positive heat gain. Overground architecture, in the meantime, is destined to move in the opposite direction — toward tighter, boxier buildings, slit-like windows and less outer skin to transfer warmth to the great outdoors.

For the moment, though, we'd do well to design living spaces with maximal floor area and minimal wall surface. In this light, a two-level house may at first look better on a chart of construction economy and energy efficiency. What may *not* appear on such a graph is the structural stress factor, which shows gigantic pressures against high vertical walls. These are so strong, in fact, that they want to push the entire second story deck out the front window (6m).

To counteract these forces, a second story should be professionally engineered, and built only of reinforced concrete. A wooden joist

Metz's own Baldtop Dugout is an example of a successful two-level underground structure.

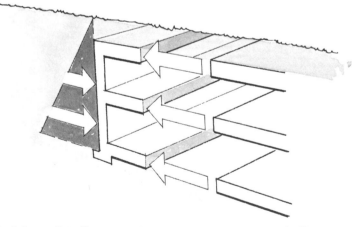

An intermediate floor must be engineered to withstand earth pressures.

system is out of the question. Further, excavation for a two-story underground house can become very expensive, and backfill material must be compacted with special care.

South-facing and westerly walls may include standard 2″×6″ studding with large window openings. The spaces between studs must be thoroughly insulated with fiberglass.

Wind currents across vast glass-covered spaces can be broken up by "fins" extending out from walls between windows. Smaller, interrupted wind patterns mean less heat loss through convection. But the fins may need "thermal breaks" (6p).

On those walls that are not to be left naked, the more earth cover the better.

Wind fins stand like sentinels along the south wall of the Winstons'. From inside they appear to be extensions of the interior partitions. The barrel-stave lawn furniture is designed by Mr. Winston, an architect.

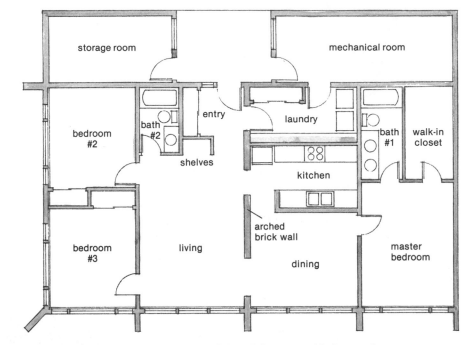

Earthtech 5 floor plan. Floors in the dining, living room, kitchen and entry are quarry tile — meant to store solar heat. A Defiant wood stove sits on the living room side of the brick arch.

6d. Floor Plans

An earth-topped roof needs to support tremendous loads — more than even a very strong truss or rafter system can hold. This high load factor (6i) demands a look at alternative roof structures such as column-and-timber systems, precast concrete planks (6aa), corrugated metal arches, culverts, silo blocks, and geodesic domes (6m).

What all this means is that a house underground probably won't contain either attic space or basement area. Furthermore, essential columns, shear walls (6j), and other internal supports limit spatial layouts and make planning more involved than usual.

A strictly elevational design — with just one "open" wall — poses an immediate threat to floor plan flexibility and challenges the creativity of the architect. The need for natural light in each room all but dictates a narrow, shallow scheme — lined-up rooms with a long hallway running east and west behind them. Again, this produces the "motel" look.

Donald Metz expands the lighting and room arrangement possibilities in Earthtech 5 simply by leaving both the south and west walls exposed. Bedrooms are at either end, while the central living, cooking, eating, and washing areas are floored with quarry tile to collect and retain heat.

At the very heart of the floor plan is the home's primary woodburning heat source — a Defiant stove that radiates warmth to all four corners of the house. The wood stove is backed up by oil-fired hot water baseboard heating units that run along the exterior walls.

If the mechanical room and storage area on the north side of Earthtech 5 were lopped off, the remaining space would amount to 1496 square feet. This is small by historical standards in American home design, but 1500 ft^2 is an area most middle-income families will have to learn to live with in the 1980s and 1990s.

6e. Storage and Mechanical Space

By using so much double-paned glass, and by doing nothing fancy with partitions, Metz brings a feeling of spaciousness to this modestly sized house. But a "mechanical room" for a boiler, hot water heater and air-conditioning units must be provided, since these things can't be tucked away in a cellar that doesn't exist. The mechanical room and additional storage area are buried at the unheated rear of Earthtech 5.

With good planning, mechanical units can also be hidden in back closet corners and disguised beneath entryway steps. And the limitless possibilities for above-ground storage shouldn't be ignored.

It's probably a waste to have an underground garage, for example. Who says cars can't be housed above grade? Inexpensive, uninsulated overbuildings can be attractive storage areas for anything that doesn't need to stay warm. The pitched roofs of these buildings can be mountings for active solar collectors that heat water. Beneath, huge, insulated hot water tanks can be buried next to the house.

6f. Entryways

Entry planning calls for special attention. An approach to living space must be recognizable and inviting. It must be open and pleasant, so there's no feeling of claustrophobia before you even get to the door. It must be separate from the more private windows of the house, and must receive people gracefully, without disturbing others already inside (2k).

A visit to Metz's Baldtop Dugout (1a, 9f) takes you briskly down stonework steps as though into a Japanese garden (2a). Guests at the Winston house, also in Lyme, New Hampshire, enter easily from the driveway, or more formally from the sunny patio outside the south window-wall. Those who experience Earthtech 5 for the first time receive an unmistakable invitation from its arched entryway on the north side.

In the Alcorn house at Cape Elizabeth, Maine, Metz married a clean-looking, above-ground garage to an underground house. The two are connected through a sloping greenhouse entryway that backs up to an insulated wall of the garage-carport.

At the Alcorns, a visitor steps out of a car, through a doorway at grade, into a solar-heated greenhouse, then walks down some bright,

One entered the Solar Hemicycle, home of Katherine and Herbert Jacobs, through the berm. Their architect, Frank Lloyd Wright, placed very small windows along the north side of the house to reduce heat loss.

Entering Don Metz's Baldtop Dugout is like stepping into another part of the garden.

plant-lined steps into the below-ground living room. This entryway solution is as charming as it is practical. Once more, a single component, in this case the greenhouse, serves several purposes — welcoming, heating, ventilating, and feeding its owners all year long.

6g. Overhang

A well-conceived earth-sheltered house is a sundial, a natural heat trap, a piece of living sculpture. Early on June 21, the summer solstice, Mrs. Oliver Winston of Lyme, New Hampshire, awoke and walked into the long corridor of her home. Directly through the window in the door at the hall's east end, she saw the rising sun. That evening, again in the hallway, she watched astounded as the sun set directly behind the glazed door at the west end of the corridor.

Her architect, neighbor, and frequent visitor, Don Metz, teases that he built a modern-day Stonehenge for her. Privately he confesses that the eerie midsummer lightway bisecting the house is neither mystical nor calculated. His main objective was simply to orient the south wall toward true south exactly (see 2b). The hallway just parallels the south wall.

What *does* need to be calculated is the length of roof overhang above south and west-facing windows. In the warm weeks following June 21, these windows must be shaded from the high angle of the sun by a "cantilever," or roof extension.

But on December 21, the day the sun is at its lowest altitude on the horizon, the entire height of the windows should receive warm solar radiation at noon. If the depth of the house is coordinated with the overhang, winter sunlight can stretch like a lazy cat all the way across the floor to the back wall.

In places like northern New England and Minnesota, a correct overhang for typical windows of full wall height is between thirty and thirty-six inches. The determining factor other than window height, of course, is the latitude of the house site.

6h. STRUCTURE

Most building codes require a roof to support thirty to fifty pounds of weight per square foot — beyond the weight of the roof structure itself. This may not be enough for earth sheltering. A cubic foot of very wet snow alone weighs as much as forty pounds.

Soil is heavier than snow: 100 to 120 lbs/ft^2 for each foot of depth. And a roof may need to hold two feet of earth or more. The need for beefing up roof structure accounts, in part, for the 10 percent increase in cost for underground construction (3g).

But there are pressures in other directions beneath the ground besides those imposed straight down by gravity (4d). More engineering expense may be incurred if you plan a very

unusual type of roof structure (6m), if you foresee long wall spans (6j), or if an intermediary floor is part of your design (6d).

6i. Loads

Experts at the Underground Space Center (1e) explain the stresses on an underground home this way in *Earth Sheltered Housing Design*:

Loads, they say, naturally make demands on structure. A "dead load" is the vertical weight placed on a structure — including the building elements themselves, any facing material, the permanent earth load, and so forth.

"Live loads" are those that may change in size, location, or direction. Snow, plants, animals, machinery, and people are examples. A tree is a live load, and when a tree in a windstorm becomes a lever arm, extra pressures are placed on a roof — from unexpected directions.

Earth-sheltered houses present a unique live load problem because the roof is concealed by soil and vegetation. It looks like regular ground surface, but heavy loads must be kept off. To find a dump truck in your bathroom one morning would be worse than awkward. The small concrete wall "tabs" that poke out above grade on Earthtech 5 serve, among other things, as transitional devices to mark where the house ends and the sideslope begins.

"Horizontal loads" on house walls are brought on by lateral earth pressure (4d). In an underground situation these are no less important than vertical loads.

In a typical dry soil, the horizontal load is thirty lbs/ft^2 for every foot of depth. When water is added, hydrostatic pressure comes into play as well, and the horizontal thrust more than doubles (see 5e). The deeper the wall, the greater the load against it. A bermed structure, then, sitting closer to the surface, is apt to have less horizontal load than one that's excavated deep into a hillside.

Horizontal loads and vertical loads work hand-in-hand, teaming up against a wall. The combination is like an immense wedge, driving downward and sideways against it.

6j. Spans

Long roof spans with inadequate support can buckle, bend, or worse, collapse. So can walls. While columns normally support vertical roof loads, in undergrade buildings, long, straight exterior walls must also be supported by "pilasters" — thicker places in a wall — or by "shear" walls. A shear wall runs at right angles to the one it buttresses, acting like a horizontal column to prevent the wall from sagging or deforming in some other way. That's one solution.

Andy Davis (9a) designed his Cave in an octagonal shape. Without the help of consulting engineers, he sensed correctly that long underground walls would be under a lot of stress from the earth outside them. So he built shorter walls — with more angles. That's another solution.

Even though computing stress loads on curved concrete shells is a complex mathematical process, it's generally accepted that it makes more sense, structurally, to design roofs and walls that are arched rather than flat. But forming and pouring curved concrete walls can be a contractor's nightmare (see 6m). Hence expensive.

After designing their own house, Mr. and Mrs. Donald Karsky skipped hiring a contractor, and went ahead with building on their own. They live in St. Croix, Wisconsin. The Karskys got around the wall stress problem by building circular walls of stacked silo blocks. That's a third solution. Since the curved blocks interlock, they need no mortar or

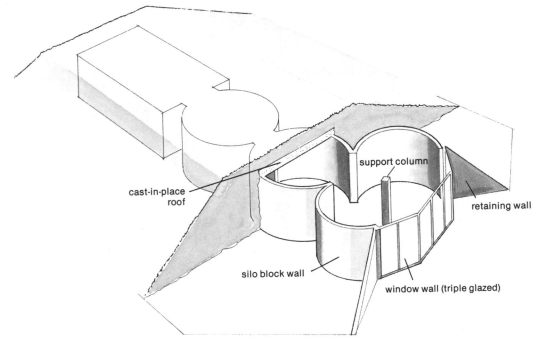

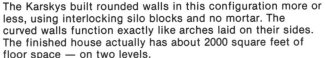

cast-in-place roof

support column

silo block wall

window wall (triple glazed)

retaining wall

The Karskys built rounded walls in this configuration more or less, using interlocking silo blocks and no mortar. The curved walls function exactly like arches laid on their sides. The finished house actually has about 2000 square feet of floor space — on two levels.

reinforcement. Once completed, the walls were connected by a cast-in-place roof.

Any walls that are tied together through a roof will brace and counterbalance each other. If they're also linked through the floor, the house becomes a box, and is stronger still. For this reason alone, floors should be in place and roofs should be installed before any backfilling is done. To be on the safe side, longer walls might be shored up temporarily at backfill time.

In an atrium-style house, earthloads are more balanced side-to-side than in a house of elevational design. The exposed elevational wall or walls can't be expected to support a lot of load, because they're primarily made of glass. In this case, the roof has to function as what a structural engineer would call a "diaphragm" —

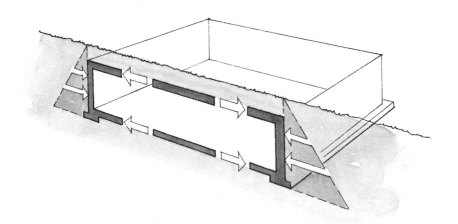

As soon as the house becomes a box, the floor and roof can act as a team.

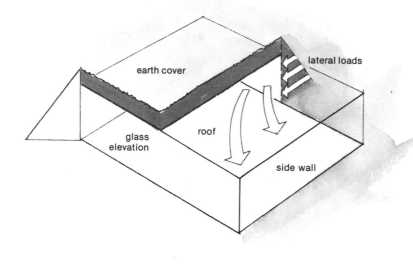

On a house of elevational design, the roof acts as a "diaphragm," redistributing the lateral earth loads to the side walls.

Where the front wall can support little, the roof can distribute the lateral loads, while the floor resists loads through friction.

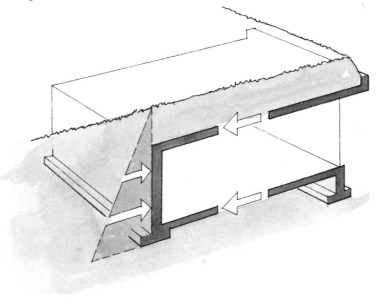

a stiff plate that strengthens the rear and side walls against lateral thrust.

Block walls, even if they're parged with fiberglass cement, need to be strengthened further with steel reinforcing bar. Pilasters should be planned along spans that are not intersected by other walls. To add more strength, wet concrete should be poured down into the openings of the blocks (6a).

6k. Footings

"Cantilevering" — building wide horizontal members at the base of a vertical wall to generate leverage or "tension" for the top of the wall — is the fourth solution, albeit expensive, to wall support (6j). A cantilevered wall needs to rest on footings that properly support the increased dead load.

Footings in underground houses are often made wider than normal, and unlike those in most conventional houses, they're sometimes reinforced with steel rod. Footing depth, of course, depends on the frostline (5d). The width of footings depends partly on the nature of the "undisturbed" subsoil at the building site (4c). In most cases, though, standard footings are perfectly adequate.

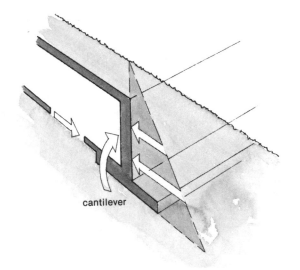

Making a wall like an inverted T or L constitutes a cantilever. If it's tied into the vertical member properly, and the footing supports it as it should, the horizontal cantilever, in combination with the floor, helps resist lateral soil pressure against the wall.

In addition to footings, Earthtech 5 has three "gradebeams," sixteen inches wide and eight inches thick, running north and south beneath the floor slab. These long concrete girders fall directly under the 2″×6″ interior partitions that help support the roof.

6l. Floors

Intermediate floors can also serve as supports for side-walls in multi-story underground buildings. That's answer number five. But because they must bear a great diaphragm load, and because of intense pressures on the high rear wall, normal wood floors are not strong enough to serve this function. They must be reinforced concrete.

Poured concrete is the strongest, and probably most economical, ground floor material (5e). About four inches must be considered minimum thickness for the slab, and this thickness must increase if there's ground water pressure (5c). Pressures from expansive clay probably can't be stopped with *any* reinforcement (4a).

Below a slab there's generally a polyethylene vapor barrier and rigid foam insulation. Steel mesh reinforces the concrete against shrinkage from within. Any type of ordinary flooring can cover the slab: slate, hardwood, linoleum, or carpeting.

6m. Alternatives

Whether they're horizontal or vertical, arches offer great strength in return for less bulk and weight. Their favorable geometry prevents loads from concentrating in one place along their length. Thus a culvert beneath a road can support many feet of earth *and* the live loads passing over it. Yet arches may be difficult to build unless premade arch supports are used. This type of curving structure will also require

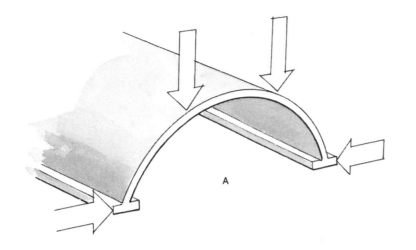

A very flat arch (A) will generate outward thrust at its base. Here there must also be some horizontal reaction inward to prevent collapse. A higher, more semi-circular arch (B) need only be supported vertically.

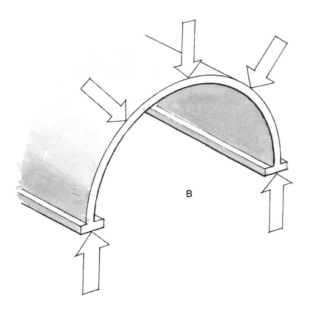

some sound engineering to be sure it's firmly anchored at both ends.

The Clark-Nelson house in River Falls, Wisconsin, was designed by architect Michael McGuire. McGuire neatly fitted the house inside two large, steel road culverts, connected the two areas, then sprayed the whole inside with polyurethane foam, meant to play a dual role as insulation and finished surface. Unfortunately the Clark-Nelson house, built in 1972, no longer meets Wisconsin's State Building Codes for insulation, but it has set underground architectural trends elsewhere.

Corrugated steel arches, such as those made by a Canadian firm called Archidome, offer an alternative to culverts. The Archidome is manufactured in two-foot panels that can be bolted together. These semicircular and parabolic structures can be as long as you want, and can be ordered in widths up to 150 feet. Strong and totally self-supporting, the arch can be insulated with rolled fiberglass, sprayed-on insulation, or synthetic foam sheets.

Corrosion of a steel arch remains a big question mark in the minds of some designers. Highway culverts can be inspected from inside, but close examination like this may not be possible in a house where insulation and wall covering have been installed. In the future,

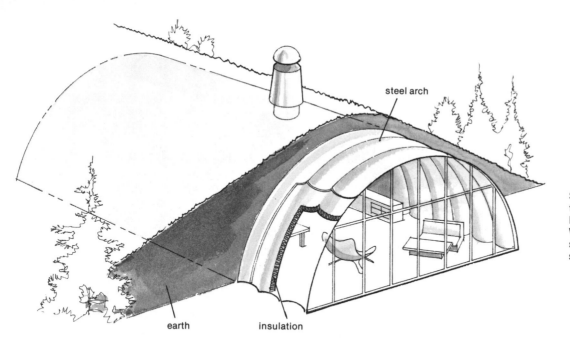

steel arch

earth

insulation

Steel arches may be used as primary supports or as temporary concrete forms for future earth-sheltered homes. Michael McGuire's Clark-Nelson house in Wisconsin used a system similar to this. The sprayed-on insulation was also the finished interior surface.

culverts may be used more as temporary concrete forms that later are taken out.

"Shells" are three-dimensional arches. The most famous shell house is the Dune House in Atlantic Beach, Florida, designed by William Morris. Its reinforced gunnite shell creates a secure-feeling, egg-shaped living space. And the view out the window toward the beach seems a little like an infant's reluctant look down a birth canal. Maybe Malcolm Wells is right again — about our wanting to go underground to return to the womb (1h).

Oddly enough, the most *popular* shell house design of all time was drawn by Wells himself in 1964. But it was never built. He calls it a Random House in a book called *Underground Designs*. Says Wells,

It's the most widely published house I've ever designed. I'd do it differently today, of course. There would be larger windows facing south, and the insulation would be on the outside for greater thermal efficiency, but I'd certainly try to keep the inviting, earthy freedom I stumbled upon here.[1]

Forming a shell is the biggest structural problem in this type of design. Wells suggests using 1/4-inch steel rods bent and tied into shape, then faced with steel wire "lath" and plaster. Other young architects in New England are experimenting with geodesic domes and inflatable balloons that can be removed once concrete has been poured and set around them.

---

1. Malcolm Wells, *Underground Designs*, p. 18.

entrance

A Random House, by Malcolm Wells

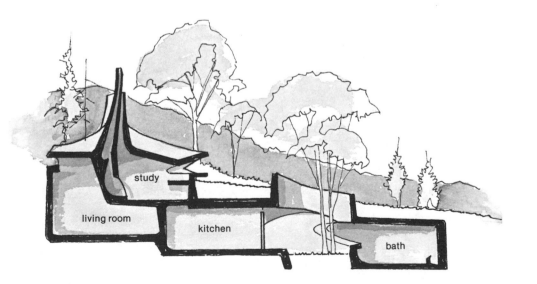

study

living room

kitchen

bath

6n. CONCRETE

Footings, floors, and grade beams will be made of poured concrete (6k, 6l). So will walls in most cases. So will roofs in *some* cases (6aa, 6bb). Strong underground walls are often made of concrete blocks (6d). Sometimes they're even made of wood — and with great success (6e). Still and all, concrete is the most widely used building material in underground construction.

Walls are normally made of "cast-in-place" concrete, meaning the wet mixture is poured into forms right on the building site. Footings and floors will also be cast in removable forms. Cast-in-place concrete has two disadvantages. It may crack if not mixed properly (5g), and residual moisture in the walls can make the house damp for months after the pour.

Steel rod reinforcing bar (called "rebar"), correct mixing techniques (6o), good waterproofing (5g), and the use of wire mesh should take care of the first problem. Long curing time nullifies the second, the wetness question. In fact, ideal planning might allow walls and slabs to be poured several months before further work is done on the house. For instance: a pour in the fall, before freezing weather but months before construction is resumed in the spring, provides more than an

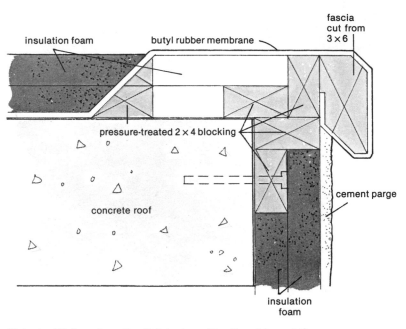

insulation foam

butyl rubber membrane

fascia cut from 3 × 6

pressure-treated 2 × 4 blocking

cement parge

concrete roof

insulation foam

Malcolm Wells solves the "tricky transition" problem at the upper elevational corner in this way.

ounce of humidity prevention — *and* a long cure.

Another way to by-pass both these problems is to use "precast" concrete that's fabricated in a plant (6r). Precast "planks," as they're called, are heavy and expensive to transport and set in place. The moisture problem is eliminated with precasting, but if some special — thereby costly — size or shape is needed, it should be ordered early. Joints between any large pieces of concrete, cast at different times, are potential leaks that must be waterproofed with extra pains.

60. Strength

Concrete has a unique quality not shared by other building materials. Instead of deteriorating with age, it gets stronger as it cures. Andy Davis is fond of pointing this out. His concrete roof bears the weight of more than three feet of soil in some places, and seems totally soundproof as a result (2l).

Davis specified 4000 psi concrete in his walls and roof. The roof is a reinforced cast-in-place slab ten inches thick. That's strong enough to support 500 lbs/ft² of load (6i).

The "psi" — pounds per square inch — designation refers to the strength of the concrete. More precisely, it describes its compression resistance. Davis's specified mixture, 4000 psi, is exceptionally strong compared to run-of-the-mill concrete. If a one-inch square device were pressed against the wall, 4000 pounds of thrust would be needed to crush the concrete at that point.

Concrete's compression strength is heightened by increasing the amount of stone and "binder" in the mixture. Portland cement is normally used to bind concrete together. The more crushed rock and the more bags of cement used, the stronger the wall. Mortar is a very weak concrete because a low percentage of cement is mixed with water and sand, instead of larger pieces of aggregate (6s).

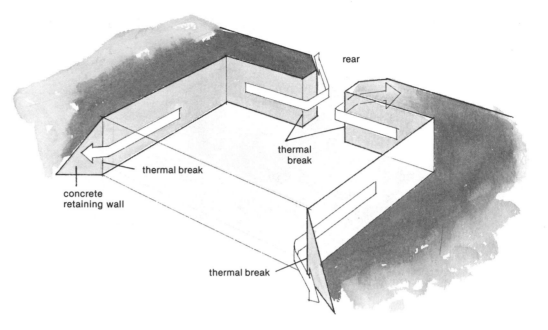

Heat from inside a house may creep along in a concrete wall and escape to the outside. A thermal break, in the form of a piece of foam insulation embedded in the concrete, can head off this heat loss. A thermal break isolates outdoor concrete structures.

rear

thermal break

concrete retaining wall

thermal break

thermal break

Like many builders, Rob Roy (9b) recommends that concrete be delivered to a job with a "stiff slump," meaning that the mixture is thick. Thinner concrete, with a higher water content, is easier to pour, but it must shed more water as it cures. This can cause added drying cracks and will humidify the house more than necessary.

Don Metz suggests 2500 psi concrete for footings and floor slabs in his houses, and 3000 psi for eight-inch concrete walls. Plans for Earthtech 5 call for a crushed stone aggregate with pieces no smaller than ¾″ to 1½″ in diameter, and rods of #4 rebar (½″) placed sixteen inches apart both horizontally and vertically.

6p. Thermal Breaks

Concrete's greatest shortcoming is that it's a fine thermal conductor. An underground wall that extends far enough to reach open air can "act like a wick" drawing heat from inside the house along its entire length to the cold outside. Styrofoam or some other insulation placed in a concrete wall at a critical transition zone can provide a "thermal break" to stop this heat flow. These insulation barriers are especially critical in a concrete roof overhang (6g), and at the junction of "wing" walls and retaining walls (6y) that connect directly to the house box.

Thermal breaks present some tricky design

problems. Blocking off the conductivity of concrete with foam insulation is easy. Unfortunately the structural continuity of the wall is usually destroyed too. The easiest way around this problem is to build noncontiguous retaining walls and roof extensions of material other than concrete.

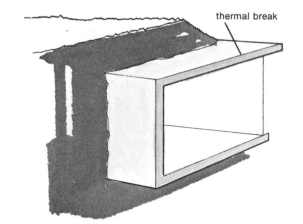

Long concrete overhangs will need to have thermal breaks to stop "energy nosebleeds," as Warren Stetzel calls them.

6q. Casting in Place

On-the-job casting of walls for an underground house involves no problems that aren't normally encountered in building conventional foundations. The forms must be braced and straightened with almost compulsive care. The walls should have a minimum thickness of eight inches — standard for most concrete walls. During the pour, the concrete should be mechanically vibrated to make the wall as bubble-free and watertight as possible. The steel reinforcing rods, which are wired together beforehand, should be in place before the first concrete truck arrives.

Sleeves for water lines, sewage lines, or for gas lines and electrical service should be placed in the forms as in a normal house. Likewise for any windows that look through the concrete.

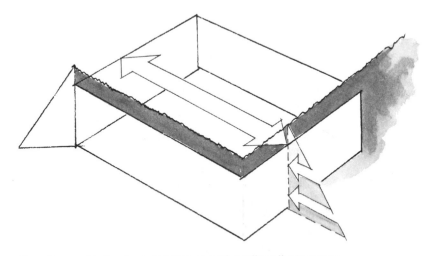

Once the roof is in place, it helps to resist sidewall pressures.

6r. Precasting

To insure good quality control and finish, precast components should be made in a factory that's certified by the Precast Concrete Institute (PCI). "Planks" vary from two feet to eight feet in width and normally have hollow channels inside. Nonstandard pieces — wider or narrower than this — have outrageously high price tags and may be difficult to piece together (6bb).

6s. BLOCK WALLS

Concrete blocks offer an alternative to poured or precast concrete. Mortar joints in any block wall are, of course, its weakest link, since mortar is basically a spacer rather than glue that binds blocks together. Mortar also provides easy passage for outside water through the tiny cracks between the mortar and the blocks themselves.

Still, Rob Roy, an economy-minded owner-builder and author of *Underground Houses: How to Build a Low-Cost Home*, used mortarless block walls in his Log-End Cave (9b), a partially buried house in Chazy, New York. He followed directions from a U.S. Department of Agriculture bulletin called "Construction with Surface Bonding."

Surface bonding is a new technique for strengthening block walls that have no mortar whatever — by coating both sides with a cement and glass-fiber parge (5h). The USDA and the University of Georgia both contend that a surface-bonded wall is six times stronger than a regular mortar-and-block structure. Walls that are surface bonded, reinforced by pilasters and steel rods at properly recommended intervals, can easily withstand normal earth pressures (4d).

As an added benefit, surface bonding is fairly good waterproofing — so long as footings below the blocks are strong enough to resist any settling that could cause the wall to crack (5g).

6t. Layup

The ratio of block wall length — or height — to thickness is 18:1. This means that in a wall one foot thick, a run can safely be eighteen feet long without an intersecting shear wall or pilaster to support it (6j). With steel rod reinforcement the ratio increases to 25:1. Since concrete blocks are usually eight inches thick, instead of twelve, pilasters will be needed every 16½ feet.

Standard spacing for steel rebar is every four feet. Beside window or door openings wider than two feet, additional steel rods will be needed. The steel is put into the hollow block cores, then "slushed" with concrete or mortar.

Interior block walls will have greater bearing strength than studwall partitions, and will add to the home's thermal mass (7a). All intersecting walls must be fastened to exterior walls with metal straps called "tie bars." Steel reinforcement will also be necessary above doors and archways. "Headers" above these openings may be steel angles or short reinforced concrete beams called "lintels."

Only the first course of blocks is laid in mortar, on top of reinforced footings. From then on, the blocks are stacked straight, level and plumb, with staggered joints. Any leveling shims or spacers needed in lieu of mortar, should be metal shims or brick ties — not wood.

As the walls are stacked, switch boxes and electrical outlets can be mounted flush with the inner surface, in openings cut out of individual blocks. Wires are then fed through the cores as the walls grow higher. "Pellet insulation," such as expanded mica or perlite, can be poured into those vertical voids not already filled with

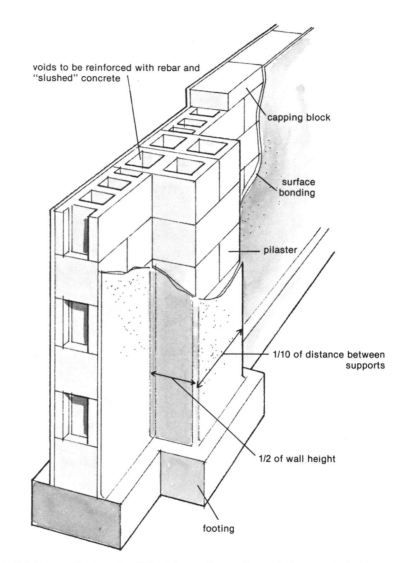

voids to be reinforced with rebar and "slushed" concrete

capping block

surface bonding

pilaster

1/10 of distance between supports

1/2 of wall height

footing

Double thickness pilasters like this, intersecting walls, and shear walls help support a long run of concrete block. They also add to the structure's thermal mass. No mortar is needed between the courses of block, but the surface bonding parge must be applied to both sides of the wall.

concrete used to encase the steel rods. These pellets can absorb moisture, in time, and settle, reducing their effectiveness as insulation.

6u. Surface Bonding

Stacked, surface-bonded walls can be erected in about half the time it takes to build the same size block and mortar wall. That estimate is straight from the USDA. The parge can be bought or concocted on the job. Ingredients in a surface-bonding mixture are water, Portland cement, hydrated lime, calcium chloride, calcium stearate, and glass fiber filaments cut into $1/2$ inch lengths. Coloring can be added as well.

Before this parge is applied with a trowel, the wall should be dampened with water. A coating $1/16$ inch thick on both sides of the wall is all that's needed. Once it's been spread, the parge can be textured with a brush or with a paint roller. If the wall is to be sanded smooth, it's best to use $1/8$ inch of bonding material in the inside surface, to insure adequate strength and waterproofing.

Surface-bonded walls can be painted with acrylic-type latex paints. And roof construction can begin as early as twenty-four hours after the bonding application.

6v. WOOD FOUNDATIONS

Just the thought of a wooden foundation is alarming to any builder who doesn't know much about them. "It'll rot!" he'll scoff. "Besides," he'll say, "the fumes'll kill you."

Pressure-treated lumber foundations are *not* a contradiction in terms. They're highly touted by the nonprofit American Plywood Association, although that's not surprising. But they're also approved by the FHA, FmHA, and HUD.

Chemicals to prevent decomposition are forced into the wood under heat and intense pressure. The impregnation reaches to the very center of each piece.

Traditional wood preservatives such as creosote and pentachlorophenol *are* toxic and may not be used legally in underground walls that enclose living space. But newer chromated copper arsenate salt preservatives are nontoxic, odor free, virtually unleachable and permanent. In other words, they're sound — and safe. They protect wood against rot and insect damage, and some even have a fire retardant.

Pressure-treated wood has the same strength, shrinkage, and swelling characteristics as regular wood. It has a soft, green hue and is somewhat heavier than ordinary lumber. Working with it requires no special tools. In some parts of the

Typical wood foundation cross-section

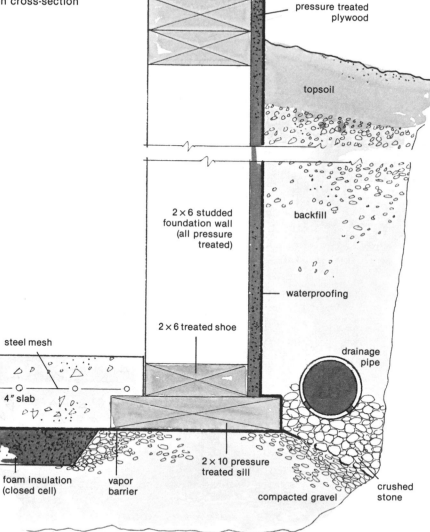

pressure treated
plywood

topsoil

2 × 6 studded
foundation wall
(all pressure
treated)

backfill

waterproofing

steel mesh

2 × 6 treated shoe

drainage
pipe

4" slab

foam insulation
(closed cell)

vapor
barrier

2 × 10 pressure
treated sill

compacted gravel

crushed
stone

country it's still a rare-enough item — and expensive enough — to require a special order through a lumber company. Any that *is* purchased should bear the FHA stamp of approval.

6w. Construction

Wood foundations become particularly attractive and practical when a wooden roof system is planned. Building a wood studwall — with its normal plywood-on-studding layout — is a process familiar to any builder. And it's easily tied into the roof.

Although it has none of the thermal advantages of concrete, a wood foundation is easy to insulate and finish on the inside. And it can be installed in any weather.

To start, a 2″ × 10″ pressure-treated sill is laid on a level bed of compacted sand or gravel. A standard studwall with plywood sheathing is then built on top of this flat sill — all out of pressure-treated lumber and rust-proof nails. A six-mil black polyethylene moisture barrier is carefully glued to the outside of the plywood foundation wall.

Backfilling is postponed until after the floor slab is poured and the roof members are all in place. Once the foundation is gently banked

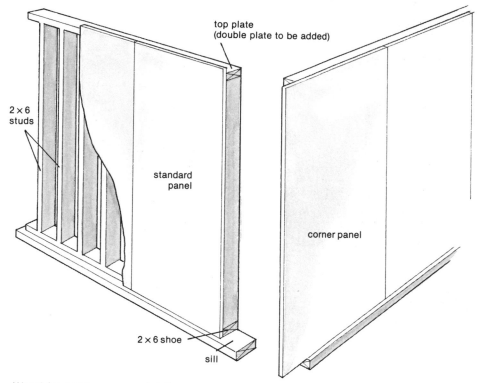

top plate
(double plate to be added)

2 × 6
studs

standard
panel

corner panel

2 × 6 shoe

sill

Wood foundation corner detail

with earth that can't puncture the polyethylene, it should be strong enough to withstand ground pressures and even sudden seismic loads that normally would crack a concrete wall.

6x. Life Expectancy

Whenever a piece of pressure-treated lumber is cut, its cut surface should be painted with more preservative. If this is done religiously, the wood should last far longer than those old non-rotting stand-bys, redwood and cedar. They haven't been around long enough to know for certain, but life expectancies of forty to eighty years for pressure-treated foundations do not seem unrealistic.

Jeff Sikora of Housing Research and Building (3g, 9e), swears by wood foundations, and will offer names and telephone numbers of satisfied customers who live either above or within them. Sikora's earth-sheltered Sub-T home (9e) does not have an earth load on its roof (6i), but its cost compares very favorably with most other designs of similar size and quality.

6y. RETAINING WALLS

Retaining walls hold back earth mounds, shape them into berms (4e), create walkways or discourage pedestrian traffic flow, and conceal private spaces (2k). Underground living areas, especially in buildings with lots of glass, are often divided from outside a window as much as by indoor partitions. This is sometimes accomplished by "fins" that continue through a glass wall (6d), and by exterior retaining walls.

These barricades can be made of many different materials: stone, poured concrete (6n), parged concrete blocks (6s), pressure-treated lumber (6v), or railroad ties. A house of

Stu Campbell

Steps are built through the retaining wall at the side of Log-End Cave.

elevational or penetrational design (6a) will generally need more retaining walls than an atrium-type house, which may require none at all. Where a concrete wing wall branches off the main part of the house to become a retaining wall, a thermal break may be needed at the connection (6p).

For gentle man-made slopes, retaining walls may not be needed at all, but can be replaced by nonbiodegradable reinforcing strips laid in the soft soil at, or just below, grade.

On the other hand, ledge at the somewhat difficult Earthtech 5 site had to be blasted to fit the house. So the retaining wall outside the master bedroom is a particularly smooth and colorful face of bedrock. In the afternoon, light

and shadow play soundless tag against the rockface protecting this small and secret outdoor extension of the bedroom.

6z. Types of Restraint

A "gravity wall" is one that's thicker at its base than it is on top. Its great weight, centered low in its mass, resists earth pressures (4d) that want to push it over. Gravity walls are often employed as retaining walls.

In cross section, a "cantilevered" retaining wall looks like an "L" or inverted "T" (6j). The horizontal members help to keep the vertical wall member vertical. Reinforcing steel in each part of the wall must be tied into the other.

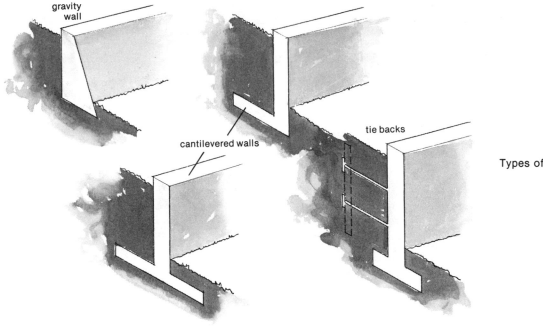

gravity wall

cantilevered walls

tie backs

Types of retaining walls

"Tie backs" are cables from a wall which attach to anchoring weights — called "deadmen" — buried within the bank of earth to be retained. These are sometimes used as extra insurance.

"Cribbing" is another typical retaining wall arrangement. Railroad ties or long concrete blocks are set into the soil to create steps. Stepped retaining walls don't need large structural members. If the steps aren't too deep, earth stress never builds too much against any part of the wall.

Seen from the side, the retaining walls reaching from Baldtop Dugout (9f) look like stylized ski jumps. They are used as much to sculpt the hilltop house as they are to integrate

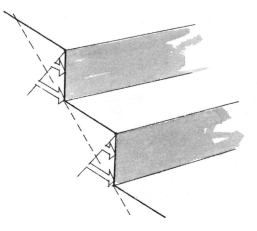

Stepped retaining walls, also known as "cribbing," maintain slope stability.

the garden in which it sits. They're made of white parged block, capped with curving precast slabs that had to be specially formed.

Soil surrounding the entryway and the northwest corner of Earthtech 5 is restrained with less visual fanfare. Here simple cribbing is done in railroad ties.

6aa. ROOFS

The worst a roof can do is hold standing water. If it doesn't pitch or drain — or both — it's a nonroof in a sense. Or eventually it will become so as it begins to leak (4g, 5a).

It must be stronger than a standard roof, able to support between 150 and 400 pounds of weight per square foot, plus a snow allowance, plus any other "live load" weight (see 6h, 6i). This limits roof design somewhat. In calculating the "dead load," which includes the earth cover (4i), weight should be figured against the saturated weight of the soil rather than its dry weight.

Whether the roof is made of heavy wood timbers like all of the Metz-designed houses, like the Davis Cave with its cast-in-place roof, or whether it's to be made of precast planks like most other underground homes, a roof should

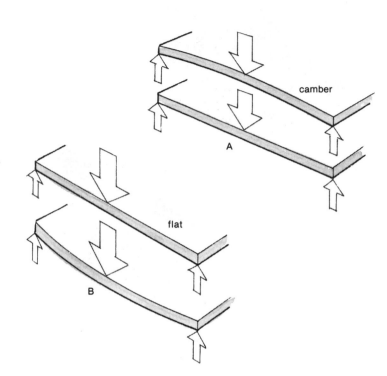

A roof with a slight upward crown (A) will flatten under a load, while an initially flat roof (B) may sag and pond.

be "cambered" slightly. Camber is a slight upward bend, such as you see in a pair of skis. With load, the camber flattens out, distributing the weight more evenly. The result should be no roof sag and no subsequent ponding (5a).

A contractor who's very experienced and fully equipped to do cast-in-place concrete roofs might build one more cheaply than someone who would raise wooden roof beams. But most suburban and rural builders will find casting in

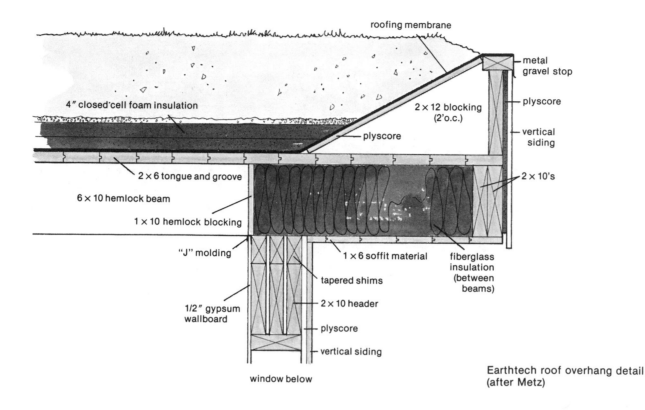

roofing membrane

metal gravel stop

4" closed-cell foam insulation

2 × 12 blocking (2'o.c.)

plyscore

plyscore

vertical siding

2 × 6 tongue and groove

2 × 10's

6 × 10 hemlock beam

1 × 10 hemlock blocking

"J" molding

1 × 6 soffit material

fiberglass insulation (between beams)

tapered shims

1/2" gypsum wallboard

2 × 10 header

plyscore

vertical siding

window below

Earthtech roof overhang detail
(after Metz)

place — with its necessarily strong but temporary substructure to support the forms — more expensive in the long run.

Besides, some find bare concrete ceilings cold and ugly. The underside of a concrete roof can be covered, of course, but the 6″×10″ roof beams in each of the homes by Don Metz, support roof planks that double as the finished ceiling. The planking and rough-sawn wood is warm, fragrant, and homey. Cut locally and

installed when green, heavy roof members for all of the Earthtech series have proved to be surprisingly economical.

He didn't do it the first time, but Andy Davis suggests building a greenhouse on the roof, to take advantage of heat rising from the chimney flue. This idea was taken one step further by John Barnard, the designer of Ecology House (9d).

In designing a home for a family in Stow,

This Stow, Massachusetts, house by John Barnard, has both an atrium and an exposed southerly
elevation. In this case, the atrium is covered with a greenhouse top.

Massachusetts, Barnard enclosed the central atrium area with a greenhouse-like cover, which acts as a passive solar collector. The owners can grow vegetables outdoors on the roof, and house plants inside the glass-covered atrium.

6bb. PRECAST ROOFS

"Precast roofs are okay," says Don Metz, "if there's an approved concrete plant in your neck of the woods that can — and will — make up planks of concrete according to your small order . . . IF these monoliths can be easily trucked to the building lot, IF you're willing to foot the cost of a crane to lift the planks off the truck and place them on the roof for you . . . and IF you're satisfied with an acoustically and aesthetically inferior ceiling . . . But all that gets to be an expensive proposition."

Andy Davis is dead set against precast roof planks, because, he claims, the weight of soil will make them bend and separate, causing cracks and leaks in the joints between the large pieces. Experience is showing that precast roofs *will* settle some, but it's difficult to know how much at this point. And it's apparent that joints between planks are endangered by water. Roof

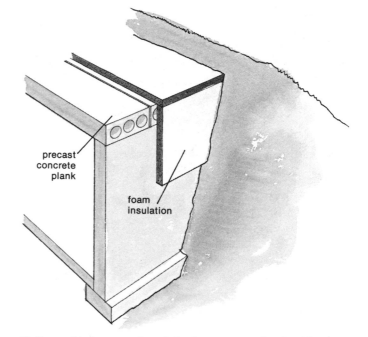

precast concrete plank

foam insulation

Hollow voids in precast roof planks may contain plumbing and wiring.

membranes can be torn if there's any sagging or shifting (5i).

Spans of more than thirty feet are impractical for most readily available precast planks — not to say unsound structurally. Twenty-five feet of

Stu Campbell

Light penetrates each room surrounding the courtyard of Ecology House, even though each is entirely covered with earth. The undersides of the precast concrete roof planks serve as an attractive ceiling.

unsupported roof should be considered the maximum safe span (see 6j). With more than eighteen to twenty-four inches of roof soil (4g), a fifteen-foot span makes more sense.

Many examples of successful precast roofs can be cited, including the low-cost, relatively fireproof Ecology Houses by John Barnard (9d). And it's safe to say that superior precast components will become cheaper and more available as time goes on. Soon, probably. This should simplify things for underground designers.

7 HEAT

7a. THERMAL MASS

As an insulator, soil is about twenty-five times *worse* than today's best insulating materials (7e). It's not earth's insulating quality, but its immense mass that makes it such an appealing cocoon for a house. The more there is of it, the greater heat change is needed, to adjust its temperature even one degree (4c). It's like a gargantuan heat battery in this respect.

It's already been suggested that throughout the summer, the sun's heat slowly warms the earth's surface, helped by tepid water that percolates downward through the soil particles (5b). As Mother Nature reverses herself and the climate turns colder, heat is retained in the soil longer than in the atmosphere. The soil surface eventually freezes, stopping any cold water flow to its lower reaches, and the earth gets covered with an insulating layer of snow. Now heat can only escape very slowly.

Great thermal mass, as many have said, stabilizes temperature — leveling off the jagged day-to-day peaks and valleys that appear on air temperature charts. In makeup, stones, bricks, and concrete are close cousins to soil, because unlike wood, or fiberglass, or Styrofoam, they offer little resistance to the comings and goings of heat. They're not so fickle as air in this respect, but in small amounts they'll accept and give back heat without putting up much fight.

Units of heat, called British Thermal Units (Btu), can be stored in large volumes of such material. And the thermal mass of a house can be increased by adding more earth, concrete, and masonry. All this storage area is worthless, of course, unless there's an energy source to "charge" the walls, floor and stonework with Btu. This heat source might be the sun, a wood burner, electric, gas, or oil-fired heaters (7b).

Water, oddly enough, stores heat better than air, earth, concrete, or anything else except a few rare salts. In fact, the whole concept of the Btu is based on the amazing heat-retaining qualities of this common substance. It takes one Btu to raise or lower the temperature of one pound of water one degree Fahrenheit.

Thermal mass, then, can also be increased by building closed containers of water — sealed barrels, tubes or culverts — into walls and masonry structures within the house. These storage cells will allocate warmth back to the living space the second they're no longer being charged. When there's excessive heat, they, and the surrounding earth, absorb warmth — cooling the living space.

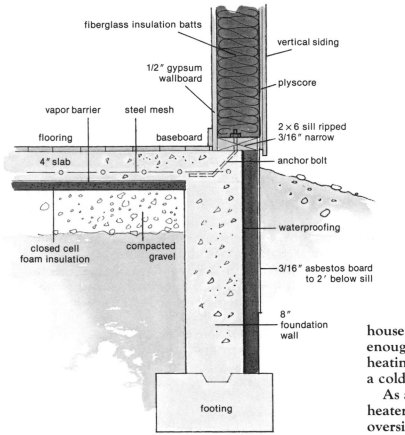

fiberglass insulation batts

vertical siding

1/2″ gypsum wallboard

plyscore

vapor barrier

steel mesh

2 × 6 sill ripped 3/16″ narrow

flooring

baseboard

4″ slab

anchor bolt

closed cell foam insulation

compacted gravel

waterproofing

3/16″ asbestos board to 2′ below sill

8″ foundation wall

footing

Earthtech elevational wall base detail (after Metz)

7b. HEATING SYSTEMS

Energy engineers Thomas Bligh and Richard Hamburger remarked publicly at the Underground Space Center in Minneapolis, "In no way can improved insulation on an above-ground building begin to compete with subsurface structures from the viewpoint of energy conservation." Indeed, a below-grade house may hold consistent temperatures well enough to make domestic hot water a greater heating requirement than space heating, even in a cold climate like Minnesota.

As a matter of fact, Bligh cautions us against heaters and appliances that are too large. An oversized furnace, for example, needs to run only intermittently if there's a low need for heat. By contrast, a small heating system, running more or less continuously, actually consumes less energy than a big one that keeps turning on and off.

Too-big cooking stoves and refrigerators can be inefficient, too. By giving off heat and moisture, they throw the temperature and humidity level of the house out of kilter (7k).

As we move underground, our concerns about warming ourselves will be modified. We'll discover, among other things, that heat is generated from sources we've never bothered to acknowledge before: from a dryer that's vented inside through a heat exchanger, from lights, a freezer, a TV, a toaster, from the oven and stove, from warm waste water and exhaust air, from pets, and even from our own bodies (1e).

7c. Passive Solar

In all but the cloudiest northern areas, a passive solar system — where no mechanical equipment is used to collect or store heat — can become a primary heat source for a house, rather than a secondary one. Well-planned southeast, south and southwest-facing windows (8d) can represent as much as 60 to 70 percent of a home's direct heat gain. Windows and skylights (8d) can be used as solar heaters during the winter — so long as there's a way to stop heat *loss* whenever there's no sun (see 7h).

The greater the ratio of south-facing glass surface to volume of living space, the easier it is to heat the structure. That's as long as the sun shines. Overheating on the sunny side of the house can actually become a problem, while back corners stay uncomfortably chilly. Fans *might* be needed from time to time to circulate air and correct this heat imbalance. But probably not.

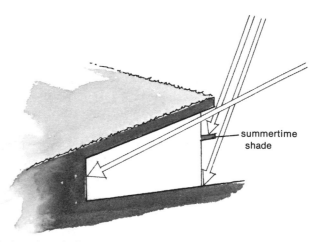

A sloped roof allows deep solar penetration in winter. However, an extra shading device may be needed in this situation.

Added thermal mass neutralizes heat imbalances. Masonry and water in walls, floors, and other heat banks absorb and redistribute excess heat. Proper ventilation helps, too (7i).

Too much glass also produces heat loss whenever the sun doesn't make an appearance. A greenhouse, connected to the main structure of the house, offers more glass to the sun, but can be closed off at night and during sunless days (6f). When it's generating heat, warmth from the greenhouse may have to be moved from the greenhouse to the rest of the home — again by fans. With plants in the greenhouse, some provision must be made to heat this uninsulated space at night. This can be done with small heating units and/or water tanks that store and reradiate heat back to the air (7a).

Active solar collectors can be considered another heat source both for domestic hot water and living space. These may call for special above-ground design features, however (6e).

7d. Back-up Systems

Because the sun can't be trusted to supply heat with perfect regularity, some conventional back-up system — or systems — will be needed. Other energy sources might be oil, natural gas, propane, coal, wood, or electricity.

Statistics on specific fuel consumption in earth-sheltered homes are still somewhat imprecise, since exact monitoring procedures have not been used until very recently. But the consensus, confirmed by whatever data are available, is that energy use of all kinds is significantly lowered below ground, even without solar input (3h). Experts also agree that more than one back-up system will be in an

Outside air reaches the wood stove in Earthtech 5 from beneath the floor.

owner's best economic interest, as various energy supplies and availabilities fluctuate.

During the earliest periods of operation, demands on all heating units may be higher than normal as long as the temperature of the thermal mass in and around the house is being raised. But after eight months or so, heating needs will stabilize.

As it looks now, a combination of sun and wood will heat underground homes in many parts of the country. Fireplaces, which are often

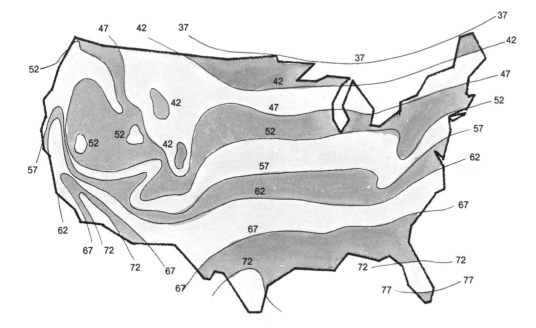

Well water temperatures as measured by Collins tend to run colder than ground temperatures in the north, warmer in the south. Still, they tend to be quite stable throughout the year, as is earth temperature well below the surface.

found to be energy drains in overbuildings, are more effective below ground. Metal wood stoves and coal burners are better, of course. Russian-type ceramic stoves, which are built into large masonry chimneys with long back-and-forth baffle systems to carry smoke, may prove best of all. Such masses of masonry can collect heat, both from the fire *and* from the sun.

Any wood- or coal-burning heater will need an air supply. What's not always considered is that an underground house with all its windows and doors closed is far more airtight than a house above grade. Don Metz and Rob Roy solved this problem in the same way — by running an air pipe from outside, beneath the floor slab to the fire, to insure an oxygen flow.

Roy computes that his 910 ft² Log-End Cave in northern upstate New York was kept comfortably warm during the winter of 1977–78 with just 3.43 cords of firewood. How does he explain this low energy consumption? He writes,

> Why is it so easy to heat a subterranean house? What is its special magic? No magic really — just common sense. . . . An above-ground structure is situated in ambient air, which in the winter reaches −20 degrees in our part of the world. If 70 degrees is required in the house, it is necessary to heat to 90 degrees above the temperature of the ambient. In an underground house, the ambient is the earth itself, about 40 degrees in the winter where we live, which means it is necessary to exceed the temperature of the ambient by only 30 degrees.[1]

A forty-degree ground temperature is about as cold as it can get anywhere in the continental United States.

---

1. Robert L. Roy, *Underground Houses*, p. 123.

7e. INSULATION

The lifeblood of any house, during cold seasons, at least, is its heat. To a designer, the "envelope" of a structure is the part where you or I live — whether it's above or below grade. An underground envelope must be insulated with artificial materials just like any other, despite the mistaken notion that soil itself is a good insulator (7a).

Earth-sheltering *does* reduce the "ventilation

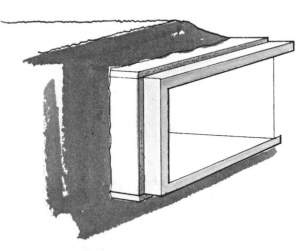

One solution is to wrap an entire concrete structure in insulation — from the outside.

load" on a house — that unwanted heat seepage in and out of the envelope through cracks and holes in its walls. Still, heat easily can be conducted out of those parts of the house that are not buried — especially through concrete, masonry, and metal. To prevent these heat veins from hemorrhaging all of the envelope's warmth, thermal breaks should be installed wherever extensions of inside walls come in contact with outside air (6p).

Insulation should wrap the entire building, including the roof, the floor, the walls, and if necessary, the roof overhang. Heat rises, of course, and as always, it's better to insulate a ceiling more heavily than a floor.

It's also better to double the insulation thickness outside the top of a wall than it is to run a single layer of insulation from slab level to the roof (7g). If a wall backs up against ledge, there should be extra insulation placed between it and the stone, so the bedrock can't suck heat from the house by conduction.

7f. Materials

The ideal insulating material for earth-sheltered construction has a life-expectancy of at least twenty years, is fire resistant, won't absorb

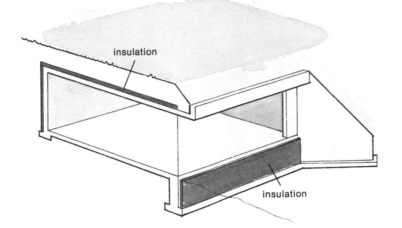

insulation

insulation

Don't forget insulation along the base of an exposed wall.

moisture, and has a high compression strength to resist backfilling and soil pressures. Ordinary rolls and batts of fiberglass building insulation won't fill this bill. But two synthetic foam boards come close.

Extruded polystyrene — commonly known as Styrofoam — is a "closed-cell" insulation foam that refuses to absorb moisture. That's why it's also used for surfboard cores and mooring buoys.

Urethane, though it insulates somewhat better than polystyrene, often has open cells that *will* receive water. Older types of urethane can soak up wetness and become waterlogged. There are also indications that urethane will degrade by as much as 20 percent over a period of five years or more. In some areas it's also attractive to ants

and other insects. Both of these synthetic materials, by the way, emit toxic fumes if and when they're ignited.

The roof of Davis Cave is insulated with 1½ inches of Styrofoam; Earthtech 5 with a total of 4 inches of "Tempchek," a closed-cell urethane board manufactured by the Celotex Corporation. The concrete sidewalls of Earthtech 5 have a 2-inch layer of "Thermax," a similar urethane product, also by Celotex.

In terms of R value — the *resistance* of insulation to heat passing through it — one inch of new urethane is rated at 8. This is better than polystyrene with a rating of R-5.4. Even with a 20 percent disintegration factor, urethane is still better at R-6.4.

Nonetheless, many designers still call for Styrofoam, possibly because it's less expensive and comes in shiplap sheets, with tongue-and-groove or square edges.

7g. Application

Underground insulation is most effective when it's placed outside the structure rather than inside. This eliminates condensation problems (7j). It can be stuck to concrete with plastic roofing cement. Both it and any

Insulation laid horizontally to a wall is even more effective than insulation against one.

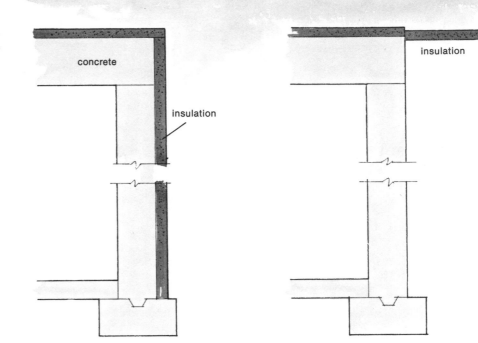

concrete

insulation

insulation

waterproof membrane outside it will need protection when the walls are backfilled. Pushing stones and gravel against the foundation can cause denting and puncture damage (5i).

Insulation need cover only the top portion of a wall — to a depth of seven or eight feet at the most. Below this level, except for the condensation question (7j), its presence is wasted because of the mass of soil above.

"Where to Insulate Earth-Protected Buildings and Existing Basements," a paper by Thomas Bligh, Paul Shipp, and George Meixel, cites several interesting heat-loss findings. Studies have confirmed the theory that below-grade insulation, laid horizontally above a wall, is more effective than an equivalent amount of the same material laid vertically *against* the wall. (See Appendix 1.) Where space allows it, this is how many underground walls may be insulated in the future.

7h. SHUTTERS

Glazing is by far the greatest source of heat loss from most houses — even without counting chilly drafts and the many Btu of heat that sneak through standard window frames at the

edges of glass. When the sun doesn't shine, windows meant to collect solar energy turn into gaping thoroughfares for escaping warmth. At times like these, glass needs to be covered with an insulating material.

Insulating drapes can be closed at night and during sunless days, of course, but some warm air will always leak around and through them. Careful fitting will help.

Appropriate Technology Corp., of Brattleboro, Vermont, is marketing its "Window Quilt," a polyester-filled drapery that runs down tracks on the sides of the windows. Various other roll-up window shades with high insulating value are also available.

Hinged insulation panels tend to take up more space than the sliding variety. One or the other should be used in an underground home, particularly on any windows on the north side.

or cloth, and it must be fitted to the window as precisely as possible. Opened, it takes up a certain amount of space, which must be planned for in advance.

Indoor shutters retain heat well, but external shutters that fit tightly against the outside of glass, offer an advantage. They allow the glass to stay at room temperature. So long as the glazing stays warm, it can't collect condensation that will turn to frost. And frost, until it melts, obstructs the flow of solar radiation when the sun comes out again. It then puddles and stains as beads of water run off the glass.

7i. VENTILATION

Skeptics worry about stale air, cooking smells, and bathroom odors in an underground house. Unfounded fears. These are removed by exhaust fans and stack vents, just as they are in any other house. Doors open, windows open, and skylights can open (8e). With just a little planning, cross-ventilation should be no problem during warm weather.

In cold weather the airtightness of the envelope might create a problem in getting fires to burn and chimneys to draw properly. This can be solved by feeding outside air to a wood-

Sun Saver, a manufacturer of thermal-shutter kits, claims that an average south-facing window with ten square feet of glass will produce as much heat during winter months as twenty gallons of fuel oil. They don't say how *much* heat is lost, but they make the documented claim that their thermal shutters, meant to be closed at night, can reduce heat loss through a standard single-glazed window by as much as 85 percent, and by as much as 73 percent in a double-glazed unit.

The Sun Saver, like most sliding or folding shutters, is designed to be installed inside the living space, so it can be operated easily. It is made of insulating foam (7f) covered with wood

Screen-filled openings tucked beneath the eaves of Rob Roy's Log-End Cave offer cross-ventilation during warm weather. When it turns cold, these are closed off with tight-fitting pieces of rigid foam insulation backing plywood.

burning unit separately — through a duct beneath the floor slab (7d). Other than this, no unique or elaborate mechanical systems should be necessary — so long as the basic design of the house conforms to standard building codes (2h).

Raised skylights or light wells in a roof, sometimes called "roof monitors," can be located above the highest point in the building. Hot air, rising along such an escape route, will draw cooler air into the house from below — even when there's no breeze. The amount of

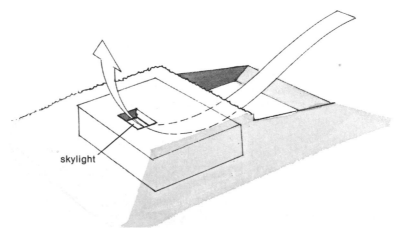

skylight

In summer, hot air rising through open skylights pulls cooler air into the house from outside the front elevation.

opening in such a roof monitor can be regulated manually. Combined with normal heat transfer to the surrounding soil, such openings should be all the cooling that's needed in all but the hottest parts of the country.

It's predicted that many homes in the twenty-first century will have "earth pipes," a maze of underground tubes in the soil near the house box. These tubes will store cool air for summer delivery, and hotter air to warm winter-chilled bones. In time, we'll have more and more direct control over the microclimates in our homes below ground. And much of this control will be nonmechanical and energy-free.

7j. Condensation

Beads of water can form on any surface that's colder than humid air next to it. In summer, cool underground walls can shrink the volume of adjacent air, causing condensation. In winter, a wet film of condensation will build and freeze on the inside of poorly insulated glass (8d).

Condensation *can* be an irritating problem if a house is kept exceptionally cool during a hot spell, then suddenly opened to a rush of warm, humid air. Thoughtful ventilation should prevent this kind of dampening event and all of its resulting damage. When it does accumulate,

condensation can be removed in one of two ways. One, let the walls warm up. As they approach the air temperature, condensation will decrease.

Two, dry the air. It's best, then, to warm an underground house with outside air on a very dry day.

7k. Humidity Control

The Underground Space Center emphasizes humidity control, too. Residual moisture in the walls, poor water and vaporproofing, washing, food-processing, cooking, cleaning, perspiration, transpiring plants, standing water, and even normal breathing — all raise the humidity level in a subterranean home. When this happens in the summer, people feel "hotter" than usual, though there may be no temperature increase.

In the house itself, metals rust, wood swells and warps, walls get moldy, and cool surfaces feel "clammy" with condensation. But potential problems like these are easy to head off. Electrical dehumidifiers may be needed to lower the moisture content in the air, particularly during the early months of occupying the house. This may be simpler, and certainly cheaper, than air conditioning, which performs the same function and lowers the temperature as well.

8 INTERIOR

8a. PSYCHOLOGICAL FACTORS

The few studies done on the psychic impact of underliving during the mid-1970s, had mixed findings. University of Texas psychologist Paul Paulus remarks,

> The occupants may have positive feelings about the uniqueness and esthetic quality of their environment and energy conservation. On the other hand, it is possible that individuals may have negative or phobic reactions to being in underground buildings. Informal interviews by Sommer (1974) suggest that most individuals react negatively to working in underground environments. They often complain of feeling like moles. Much of the reason for this may be the lack of windows. . . .[1]

Frank Lutz, a professor of education at Penn State, headed a research team from January 1963 to January 1964, which found out something different. They studied the effects of the earth-sheltered Abo Elementary School in Artesia, New Mexico, on its students.

In both this and in a follow-up study in 1971–72, the group concluded that being underground had no adverse effect on the achievement or on the physical and mental health of the pupils:

> Our original study concluded that the Abo Elementary School and Fallout Shelter was operating as an effective elementary school environment. Our present study confirms this finding. . . . There is no evidence that the fact that it is underground and serves as a fallout shelter and is so marked, designated, and named, has had any detrimental effect on pupils attending that school. . . . Apparently other schools could be so constructed and if operated under similar conditions, would also serve as effective elementary schools.[2]

Since that time, of course, many more underground schools and factories *have* been built — almost all of them with plenty of windows. Many have received national publicity. Educators, psychologists, and business people are recognizing that the physical and psychological well-being of an underbuilding's inhabitants depends directly on the quality of its design.

---

1. Paul Paulus, "On the Psychology of Earth Covered Buildings." *The Use of Earth Covered Buildings*, p. 66.

2. Frank Lutz, "Studies of Children in an Underground School," Ibid, pp. 75–76.

In his almost mystical wisdom, Frank Lloyd Wright recognized that the sunny side of a partially buried house could have vast areas of glass to complement its tremendous thermal mass. Stone walls and concrete floors were charged with solar radiation during the day. At night drapes could be drawn, and heat given back to the living area. As in all Wright designs, great attention was paid to functional detail — right down to the furnishings.

Even today, residents and workers in the subterranean section of downtown Montreal have no reason whatever to go above ground on a cold winter day. Some never go outdoors for weeks at a time. Some members of the next generation will never think twice about spending as much time below the earth as they do on its surface. It's a matter of conditioning (1b).

8b. Visual Stimulation

Psychologist Paulus goes on to say, "Individuals in windowless environments often complain of boredom due to the lack of visual stimulation." It almost goes without saying that an essential ingredient in any underground design must be windows that offer both light and a view of the outside world. That's unless the house is *meant* to be a bomb shelter (1d).

In those parts of the house with no windows, "window surrogates" may be needed. These might not be the fake window-murals done by bomb shelter–home designer, Jay Swayze (1d), but he had the right idea. Fish tanks, sculptures, mobiles, terrariums, posters, wall textures, color schemes, and subtleties in artificial lighting can all heighten the interest of a home's inside. The

underground housing boom will surely give birth to a whole new breed of interior decorators.

Light, especially natural daylight, will continue to be the most powerful stimulus of all. Quality and direction of lighting are critical, as Don Metz says, to "balancing the ambience" indoors. The greater the variety of light sources, the less the chance of our feeling buried alive.

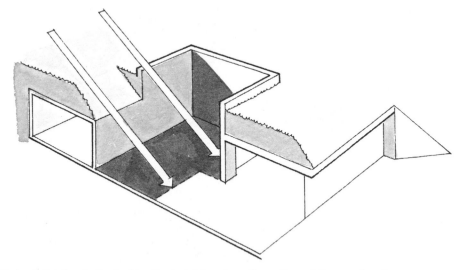

Natural lighting is limited by the height and configuration of the courtyard.

Marilyn Makepeace

The inside of Baldtop Dugout is a truce between straight lines and curving surfaces. Level changes and contrasting textures add personality to the well-lit but seemingly dimensionless spaces. The cat, by the way, is alive. The birds are sculpture.

Living in a culvert that opens at only one end (6m) could become unnerving after just a few days.

Light penetrates Metz's Winston house from the entire south wall, and through doorways at the east and west ends (6g). The dining room, farthest from the window wall, sits directly beneath a light well. A shower of light, reflected off angled ceilings in this cupola, pours into this innermost section of the house at all hours of the day.

Baldtop Dugout, nestled comfortably into its hilltop with nearly 360 degrees of view, receives light from all sides. Earthtech 5, open on two walls, has an additional light source from above its rear entryway on the north side (6f). The darkest part of Earthtech 5 would be the back corner of the kitchen, were it not flooded with artificial lighting. Even so, Metz almost wishes he'd put a skylight there.

8c. Decorating

The interior of an earth-sheltered house must be dry, bright colored, and well ventilated. High ceilings, growing plants, and see-through space dividers reduce any sense of closeness, crowding, or isolation. Light-colored walls reflect and enhance light, while dark walls absorb it. A clever mixture of surface colors and different textures on walls, ceilings, and floors can establish contrasting moods in spaces immediately adjacent to one another.

An unimaginative underground box can feel like a tomb. But a sensitive juggling of spatial proportions brings a house to life. Changes in ceiling height might be difficult in a below-grade building, but simple level changes in a floor take us quite dramatically from compressed space to expanded space and back again. Boxes have square corners and definite boundaries. Curved spaces, by contrast, can seem dimensionless (1a).

Rounded walls in Baldtop Dugout lure a new guest toward the home's communal areas. Straight ones issue a gentle warning that they border more private spaces. The arched masonry wall in Earthtech 5 breaks up the squareness of the main living-dining area. It's a partition or not, depending on the light adjustment.

By equalizing light, a hostess in the kitchen can see and be part of any guest activity in the living room. Later, this same fire-warmed wall may seclude the adjoining room, standing guard for an intimate candlelit dinner.

8d. WINDOWS

Generally speaking, north-facing windows almost always produce negative heat flow — away from the house — while south-facing windows are heat gainers (7e). In fact, a window on the north side of a building may lose twenty to thirty-five times as much heat as an insulated section of underground wall that's the same size. East- and west-facing windows, when there's no sunshine, will lose approximately ten times as much heat as the wall. This seems reason enough to put big windows on the south side, and very small windows on the north.

Some of us will at first find this contrary to tradition and taste. Artists have always preferred to work in north light. Above the equator, views through northerly windows are usually well lighted and always glareless. If scenery to the north is spectacular, we'll resent turning our backs on it by opening the house only to the south. When visibility through one whole side of a house is cut off, we can feel paranoid about what's going on out there. Besides, we sometimes need cross-ventilation (7i).

As the fossil fuel famine gets worse, we may find ourselves looking through tiny window openings. But have you noticed how small

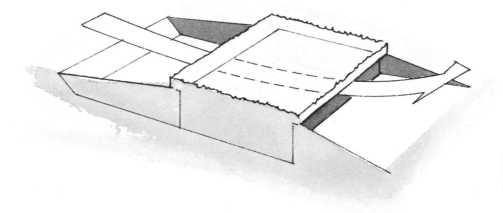

Small windows on a house's north side will lose some heat, but they'll provide pretty light and guarantee cross-ventilation.

windows seem to get larger as we move closer to them?

Malcolm Wells (1h) reminds us that we often see the world through small openings — car windows, for example — and that a view through a little window carefully placed at eye level can be far more precious than the huge panorama seen through a glass wall. Considerate architects in the late twentieth century will know this, and will provide seating places close to small windows.

All windows will want multiple glazing, naturally, and should have insulated drapes inside, or shutters that close from *outside* the glass (7h). Like all solar-heated homes, the underground design must consider the position of the sun at various times during the day and its altitude above the horizon at all times of the year. This will have a profound effect on window height and on the length of a roof overhang (6g).

A two-story house with an atrium (6b) will need about twice as much courtyard space and almost double the glass if sufficient natural light is to enter the home. A sloping roof on an elevational house lets low winter light deeper into its rooms. But if the slope gets too steep, the roof will erode (4g).

Windows, on the other hand, must never slope, but should be installed vertically. Tilted windows lose far more heat than they should. Because the bottom of the unit is heated more than the top, a circulating air current is set up between the two layers of glass. This circulatory movement, known as *convection*, automatically increases heat transfer.

8e. SKYLIGHTS

Skylights offer a solution to building code requirements insisting that each habitable room be supplied with natural lighting (2h). Again, it should be understood that the skylight is *an* answer, though not always the best one. In most cases, it represents a serious wound in the

Phokion Karas

A glass-covered atrium, like this one by John Barnard, creates a solarium in the very midst of the living space.

house's heat-retaining capacity (7e). Most energy-conscious designers frown on it for just this reason. The priorities of light and heat must be carefully weighed.

Nonliving spaces such as bathrooms, closets, and other storage areas should be lighted artificially. Flat or bubble-type skylights, although recently popular on above-ground homes, should be used sparingly. Don Metz's own house has a flat skylight above the kitchen, but the lone skylight piercing the roof of Earthtech 5, a later design, does so over the outside entryway where there can be no heat loss.

When they're placed over a heated part of an underground house, skylights should be massively insulated, carefully flashed and waterproofed. A leakage problem in Rob Roy's Log-End Cave (9b) happened right next to the skylight. It was readily solved by improving the drainage on that part of the roof. Excellent, though expensive, insulated and watertight units are now manufactured in Scandinavia and are gradually being imitated in this country.

A skylight for remote corners of a subterranean home should be the "operating" type, meaning that it can be opened to serve as a ventilator or fire-escape route if need be. Adding rope ladders and built-in stepping places,

Marilyn Makepeace

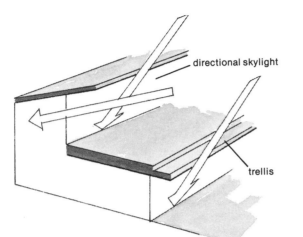

A trellis shades only in summer. Directional skylights, with vertical glass, lose less heat than flat ones.

in combination with smoke detectors, should satisfy safety regulations, FHA minimum property standards, and most local building codes.

"Directional" skylights, which face south, east, or west, use vertical rather than horizontal glazing and are better heat preservers. These periscope-like windows are easier to insulate, can be effective heaters in winter, and can be shaded by an overhang to function only as vents and light sources during the summer.

The "light well" in the Winston House is actually an east-facing and a west-facing directional skylight placed back-to-back (8b).

Except for the chimneys and "light well," the roof of the Winston House is impossible to distinguish from the hillside field in which it sits. The light well — back-to-back directional skylights, actually — brightens the dining room at the back of the house. Light is reflected off angled ceiling planes, so the rearmost portion of the building is well lighted at all times of day.

8f. PLUMBING AND WIRING

All plumbing, heating, and wiring must naturally conform to local building code requirements (2h). Beyond that, an underground house issues no special challenges in terms of mechanicals (6g).

The hollow channels in precast roof planks (6bb) make ideal routes for wires and plumbing tubes. Pipes and wiring can also run through "chases" — small tunnels in a wall or concrete floor slab.

Andy Davis, mentioned so often in earlier chapters, left 6"×6" trenches in his concrete floor. Here he ran plumbing and electrical circuits. (Most of the electrical switches in the Davis Cave are on the floor — meant to be operated with the feet.) These recesses are covered with pieces of ¾-inch plywood, which lie flush with the top of the floor slab. Carpeting then covers everything.

Chases within concrete are also made of four- to six-inch cast iron or PVC plastic drain pipe. Wiring that runs through them may be placed in "conduit" — smaller pipes that encase and further protect electrical circuits.

The walls of Earthtech 5 are covered with standard Sheetrock nailed to 1"×3" wood strapping fastened to the concrete. Don Metz had to choose between leaving the concrete exposed — which has thermal advantages — or insulating the inside of the house from its own heat banks because of this easy-to-finish wall covering (7a). For the sake of versatility and familiar appearance, he opted for the half-inch gypsum wall board so common in modern homes.

The ¾-inch space between the concrete and the Sheetrock also allows room for wiring — an important factor in Metz's decision. Piping for the second bathroom and for hot-water baseboard units in the west-side bedrooms travels through a boxed-in chase passing inconspicuously over the main entryway door.

With foresight and planning, wiring, plumbing, and even junction boxes can be placed in concrete forms before a wall is poured (6q). Block walls can be wired as they're being laid up (6t). Any pipes to be embedded in concrete, whether they're to transport water or to enclose wires, should *not* be of aluminum, which will quickly corrode. Most plumbing is either copper or plastic. Most electrical conduit is cold-rolled steel. None of these materials presents a problem in concrete.

Plumbing cannot be run inside a cold, unsheltered or uninsulated wall, of course, or it will freeze. It's good practice to isolate any

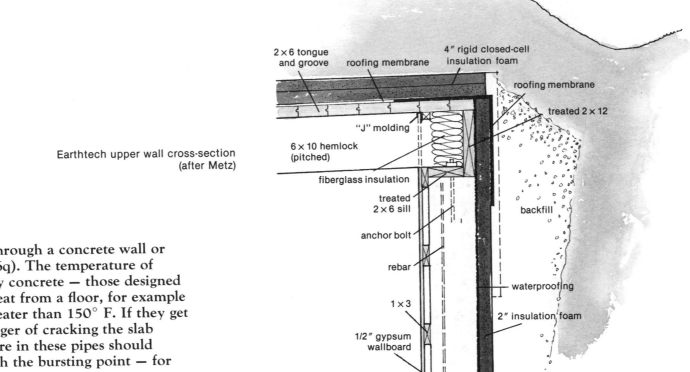

2 × 6 tongue and groove

roofing membrane

4″ rigid closed-cell insulation foam

roofing membrane

treated 2 × 12

"J" molding

6 × 10 hemlock (pitched)

fiberglass insulation

treated 2 × 6 sill

anchor bolt

rebar

1 × 3

1/2″ gypsum wallboard

backfill

waterproofing

2″ insulation foam

Earthtech upper wall cross-section (after Metz)

piping that passes through a concrete wall or slab with a sleeve (6q). The temperature of pipes surrounded by concrete — those designed to supply radiant heat from a floor, for example — should not be greater than 150° F. If they get too hot, there's danger of cracking the slab nearby. And pressure in these pipes should never even approach the bursting point — for obvious reasons.

9 PEOPLE

9a. ANDY DAVIS

He knew of no other house models to study, but two things prompted Andy Davis of Armington, Illinois, to go underground. First, he lived in a drafty overhome where heating bills during the cold Illinois winter ran as high as $150 a month. Enough said.

But his second inspiration came in the intense heat of an Arkansas summer afternoon, when he discovered the cool inside of an abandoned mining tunnel he was visiting. The deeper he went, the more comfortable it became. And he was so intrigued by the reflected beam of his flashlight on the rock walls, he let this fascination influence the interior of the Davis Cave, where all walls are faced with natural stones, set in concrete.

As the Cave began to take shape in his mind, Davis kept hinting to his family that they should be living below ground. In time, his wife began to share his enthusiasm, but the children resisted, afraid they'd be nicknamed "the Flintstones" by neighbors and friends. Finally they relented to their father's persistence, and agreed to help.

Since it was completed in 1975, the Davis Cave has received world-wide acclaim, much to the family's surprise. It cost $15,000 to build and furnish, including the land, septic system, carpeting, even new appliances and furniture. With 1200 square feet of floor space, the $15,000 bottom line breaks down to $12.50/ft^2 for the original eight-sided structure (3g). Later a rectangular addition was put on. Keep in mind, however, that the Davis family did all of the work themselves.

The design proved so successful and has attracted so much national publicity that Davis has begun to franchise the idea, selling plans and construction concepts to building contractors for a $3900 fee that guarantees an "exclusive territory." He claims that homes similar to his own can be built for 5 to 10 percent below conventional construction costs, but adds, "The big savings start when you live in the home" (3h). Various plans, available through the Davis Cave Company, have names like Dreamstone, Logstone, and Spanishstone.

The original Davis Cave has cast-in-place walls and roof (6q), two large round windows that face west (8d), door knobs made of cow bones, and animal fur on the doors. The different rocks and stones embedded in the walls not only accent the primitive decor, they increase the home's thermal mass (7a), taking a

Davis Cave, Armington, Illinois

long time to heat up and cool down. Passive solar heating was not a conscious part of the design.

A Franklin stove is the primary heat source, and it consumes roughly 2½ cords of firewood a year. His oft-repeated story about running out of firewood during a subzero cold snap has helped make Andy Davis a household name among underlivers — both real and prospective. He says,

We lost only 2 degrees a day. It was something like 25 degrees below zero outside, with a wind chill factor way down to minus 80 degrees, but in four days without a fire the temperature in here only slid down from 70 to 62. To tell the truth, I could have gotten out and brought wood in a day or two earlier than I finally did . . . but we were all so comfortable that I didn't want to do it. We just didn't have anything to worry about.[1]

1. Underground Space Center, *Earth Sheltered Housing Design*, p. 223.

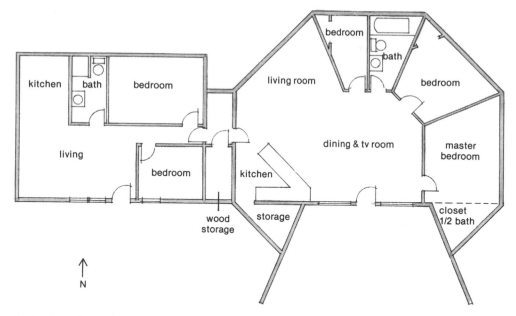

Davis Cave, floor plan

Like Rob Roy and others, Davis points out that the temperature differential between the earth (usually 55 degrees) and room temperature (70 degrees) is far less than the difference between room temperature and the outside air — which can go well below zero. Far fewer Btu are needed to raise the temperature in living space 15 degrees than to raise it by 90 degrees (7b). This is why Davis chooses to describe his homes as "earth powered."

In a letter to me, Andy writes, "The three years that my family and I have lived in our Cave have been the happiest years of my life." In his down-to-earth way, he's encouraging all of us to follow suit by choosing what he calls the underground "life-style option."

The home of "Caveman" Andy Davis features fur-lined doors, wall stones embedded in concrete, and foot-operated electrical switches in the floor.

Log-End Cave, Murtagh Hill, West Chazy, New York

9b. ROB ROY

Rob Roy, as his name hints, is an architectural rebel. He's a designer-contractor who objects to building codes that raise housing costs and keep many owners from doing their own construction. "Man has the right to build his own shelter," he insists. "Just as he has the right to starve if he chooses."

The Roys are independent. Rob, wife Jackie, and young son Rowan, live in Log-End Cave, dug into the side of Murtagh Hill, an alternative community near West Chazy, New York. Murtagh Hill has no electricity except what can be generated from the wind. And the residents want to keep it that way. "Self sufficiency" is a common phrase up there. To Rob Roy the idea of living in the earth was as natural as growing his own food.

Log-End Cave encompasses about 900 square feet of living space, and was built completely for less than $8000 — partly from scrounged and recycled materials. Roy describes, step-by-step, the process of building the cave in his friendly book, *Underground Houses: How to Build a Low-Cost Home.*

His home is rustic, but in no way "funky" as the use of old silo hoops, glass bottles, and used

beams might suggest. Massive stone retaining walls (6y) and heavy wood timbers characterize the structure and give it a medieval flavor.

The south face of the house — oriented from the North Star — is made of thermal window glass set in a Tudor-like wall of "stove-wood" masonry. The Canadian technique of combining carefully cured sticks of firewood, mortar, and fiberglass insulation to fashion a warm and attractive wall is discussed in an earlier book by Roy, *How to Build Log-End Houses.*

Passive solar heating (7c), heat storage in both the walls and a masonry heat "sink" in the center of the living area (7a), allow the house's

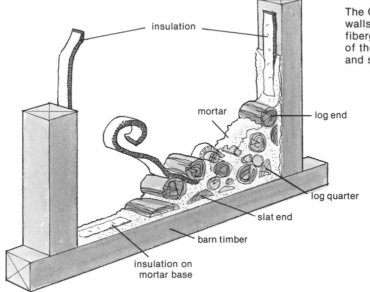

insulation

mortar

log end

log quarter

slat end

barn timber

insulation on
mortar base

The Canadian technique for building log-end
walls is used by the Roys. Thin strips of
fiberglass insulation are laid in the very middle
of the wall, covered with log ends cut to length,
and surrounded with mortar, both inside and out.

Stu Campbell

The room just inside the Roy's door is painted white and is
used primarily to store firewood. It also functions as a sort
of airlock. Light comes through the window, as well as
through the bottles mortared into the log-end wall.

large cookstove, its only necessary heat source,
to be fueled with less than 3½ cords of
firewood per winter (see 7d). This is how the
Roys live comfortably on very little cash
income.

"Surface bonding (6u) and plank-and-beam
roofing unlock underground housing for the
owner-builder," says Roy with some intensity.
All but the south wall are made of concrete
blocks laid up without mortar. Special glass-fiber
cement was then applied to both sides of the
wall to bond the blocks to each other.

Since this strengthening parge is not entirely
waterproof (5g), Roy spread plastic roofing

cement and applied a six-mil polyethylene membrane to the outside of the walls. Later foam insulation was placed outside the waterproofing.

The roof of Log-End Cave is hemlock timbers and 1½-inch tongue-and-groove planking, beneath two layers of Styrofoam insulation. Roy also used black polyethylene on the roof, covered by several layers of roofing paper and tar. The "poly" membrane needed protection from stones, so the soil directly above it was sifted first. This also provided good roof drainage — for the most part.

Few problems have plagued Log-End Cave since it was finished in early 1978. There *was* the skylight that leaked, but was easily fixed by improving the drainage next to it. Then hydrostatic pressure combined with earth pressure to threaten the wide fascia board on the front of the roof. Redraining the roof solved this problem, too. They wish they'd placed an inch of insulation under the floor slab, and put insulation lower on the walls (7g). And if he had it to do over again, Rob would run the subfloor stove vent (7i) straight out through the base of the wall rather than jogging it past the footing where it collects moisture. Otherwise they're content.

Inquiries from potential customers for other Log-End Caves are flowing into Murtagh Hill, home of back-to-the-land underliving. "We've made a few mistakes," says Roy, "but nothing serious. And we're learning a lot about how people can build inexpensive, energy-efficient houses for themselves."

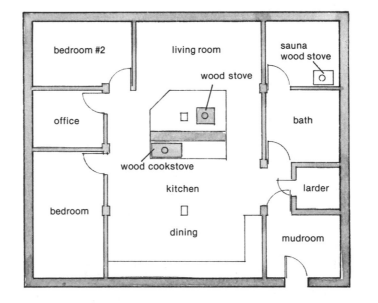

Log-End Cave, floor plan

9c. MALCOLM WELLS

It would be impossible to omit Malcolm Wells from *any* discussion of earth-sheltered building (2l). The Titan of terratects, Wells is clearly the banner-bearer in the crusade for underliving, and has been for some time. *Progressive*

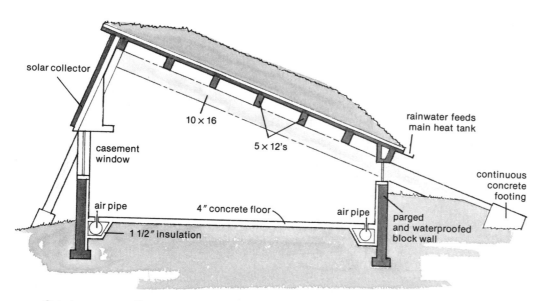

solar collector

10 × 16

5 × 12's

casement window

air pipe   4″ concrete floor   air pipe

1 1/2″ insulation

rainwater feeds main heat tank

continuous concrete footing

parged and waterproofed block wall

Solaria, cross-section

Architecture published an article by Wells in 1965 called "Nowhere to Go But Down." It appeared a year after he designed the romantic hobbit hole he named a Random House (see 6m). He wrote,

> . . . What we do in the name of architecture is just as ruthless, just as destructive as the work the buffalo hunters did. . . . The simple fact remains, though, that there isn't a building as beautiful, or as appropriate, or as important as the bit of forest it replaces.[2]

Strong words. And effective.

About ten years later, his famous house, Solaria, was built in Indian Mills, New Jersey, a suburb of Philadelphia. In cross-section, Solaria appears to be a wedge driven back into the earth. "The house is based on a design theme so simple," says Wells. "We've done dozens of variations on it, and dozens more are possible."

The rear point of the triangular shape is snipped off, allowing a short north wall with small windows for natural lighting (8d) and cross-ventilation (7i). The main 10″ × 16″ wood timbers extend through the rear wall to a

---

2. Malcolm Wells, "Nowhere to Go But Down," *Progressive Architecture*, p. 176.

Solaria, Indian Mills, New Jersey
(Front view)

continuous concrete footing buried behind the house.

The short walls front and rear, like those at either end, are concrete blocks parged with waterproof cement (5g). The roof was an early and successful experiment with butyl rubber sheeting wrapped completely around a beveled wooden drip edge. This waterproof membrane covers 1½ inches of rigid foam insulation that rests on the tongue-and-groove roof decking.

The front wall of this semi-elevational design holds casement windows for passive solar heat, but it's also a mount for actively operated solar collectors. Within these Thomason-type collectors, water trickles down blackened metal sheets enclosed in glass. Once it's heated against the metal, the water is then transported to other parts of the house. This, combined with the two feet of mulched sod cover on the roof, has made Solaria a prototype for a whole generation of sun-heated, earth-sheltered homes.

Beyond that, Solaria is billed in *Earth Sheltered Housing Design* (1e) as "a serious effort to incorporate energy efficiency with the aesthetic aspect of earth sheltering." Energy efficiency is there all right, but in terms of aesthetics, Solaria, long, narrow, and functional as it is, can hardly be considered a sculptural

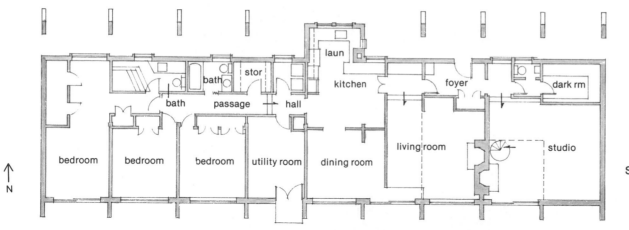

Solaria, floor plan

Robert Homan

Solaria, Indian Mills, New Jersey (rear view)

treasure. Nor is it a piece of work that's entirely "beautiful, or as appropriate, or as important as the bit of forest it replaces."

The interior is open, bright, and lovingly decorated beneath its stained wooden roof girders. And the exterior is exquisitely landscaped. But none of this disguises a front elevation that's nondescript at best.

Further, Wells's 2800-square-feet floor plan succumbs to the rooms-in-line pragmatism usually associated with elevational designs (6d). The monotony of the bedroom wing is broken by a floor level variation in the passageway. The living room and studio at the opposite end of the house are also sunk below the foyer level (8c).

Any criticism leveled at Solaria's design can only be made in light of still more successful and visually pleasing solutions done since its completion. And this in no way detracts from Malcolm Wells's monumental contribution to subterranean architecture. Wells has moved beyond it, and Solaria remains a significant milestone in this field.

The entryway to Ecology House passes over part of the courtyard's natural drainage system.

9d. JOHN BARNARD, JR.

He's built elevational style houses too, but John Barnard's name is usually associated with open-courtyard, atrium-type designs suitable for flat building lots (6b). The Ecology House series evolved from the architect's own 1200-square-foot vacation home, built in 1973.

By 1979 trees and other vegetation on the original Osterville, Massachusetts, site, not far from Barnard's office, have grown enough to almost hide the first Ecology House (2a). That's exactly as it should be, since Barnard's thesis is that space on a small lot should be used to grow things — not be covered with buildings. His houses, he says, are designed "to resemble a park rather than the ugly sprawl that now blights much of the country." In short, he equates building underground with better land usage (1e).

John Barnard's well-known claim that his underground houses save the homeowner 25 percent on construction costs (3g), 75 percent on energy costs (3h), and almost 100 percent on maintenance expenses (3f), has long been used as ammunition by those arguing the case for underliving. But Barnard makes a special point of emphasizing the need for planning, pointing out that once there's commitment to design and construction, changes become extraordinarily difficult and costly (6a).

Ecology House is covered with ten to sixteen inches of soil. This earth cover is supported by steel beams and precast roof planks (6bb), waterproofed by a built-up roof (5i) consisting

Ecology House floor plan

Labels in floor plan: utility room, kitchen & dining, living room, bath, open atrium, bedroom

of three layers of sixty-pound roofing paper and tar. The walls are cast-in-place concrete (6q) waterproofed with mopped hot pitch. As in Sikora's SubT (9e), no specially skilled craftsmen are needed to erect the simple shell.

Barnard includes air conditioning and humidity control in his design package. The first Ecology House also featured an active, ground-mounted solar collector as a back-up system for a forced hot air furnace (7b). All chimneys, ducts, and vents pass through the side walls instead of the roof. This is to prevent water from leaking in and hot air from leaking *out*.

The central courtyard, unless it's covered like one in a later Barnard house in Stow, Massachusetts (6aa), can have problems with

wind turbulence and swirling snow. The atrium can also cause serious wetness problems, especially if the house is excavated into a site with poor soil percolation (2f). So a well-planned natural drainage system should be devised to keep the courtyard dry.

Barnard is such a veteran in the comparatively new age of underground building that his design concepts are almost taken for granted. One gets the feeling that young experts in the field often quote him without realizing it. Barnard was among the first, for instance, to discourage the use of sump pumps in lieu of gravity for removing underground water from an atrium (5f).

He cites a third difficulty: an atrium is, after all, a hole in the ground that people and animals can fall into. This "pitfall" in design is easily neutralized by fences and landscaping barriers — either artificial or growing (4l).

Frank and sensible designs like Barnard's will dispel negative attitudes about underground houses (1b). Editor Roger Griffith's trip to Ecology House early in 1979 did much to inspire the writing of this book. Roger's wife began the visit convinced she would hate it. She came away a convert — like hundreds of others who have seen Ecology House's bright interior. Case in point.

9e. JEFF SIKORA

Housing Research and Building, a construction firm in Swanton, Vermont, specializes in single-family, low- to middle-income houses. Jeff Sikora, chief designer for HRB, seems as much an altruist as practical builder. His appealing designs for small, inexpensive, earth-bermed homes make clever use of space, standard building materials, and simple construction techniques. For instance:

Sikora's clean-lined SubT has less than 1100 ft^2 of floor space, but it's comfortable and well lighted. Besides, it has a low 1979 price-tag of about $33,000, not including excavation costs (3g). And it's approved by the FHA. SubT may be small, but there's abundant storage area in the attic beneath the salt-box roof. The roof, in this case, is not covered with soil.

The philosophy at HRB is that underground building costs should be comparable with typical, one-family, ranch-style construction. Actually Sikora, "father" of SubT, might object to our use of the word "underground." He makes a distinction between an "underground" home, which has earth on the roof, and a "subterranean" home, which has an exposed,

but extremely well-insulated roof. SubT is thus classified as "subterranean."

The one-story house would have a southern orientation (2b), a wood foundation (6v), a shingled roof framed with ready-made trusses, a

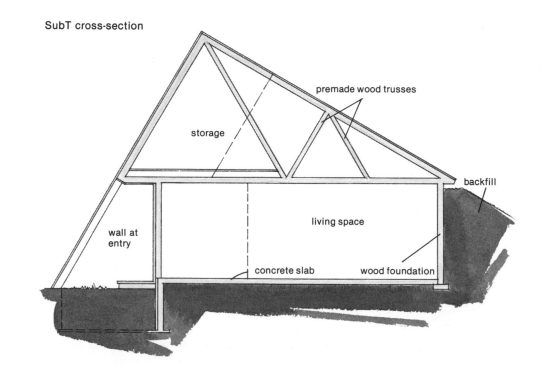

SubT cross-section

storage

premade wood trusses

backfill

living space

wall at entry

concrete slab

wood foundation

SubT, Swanton, Vermont

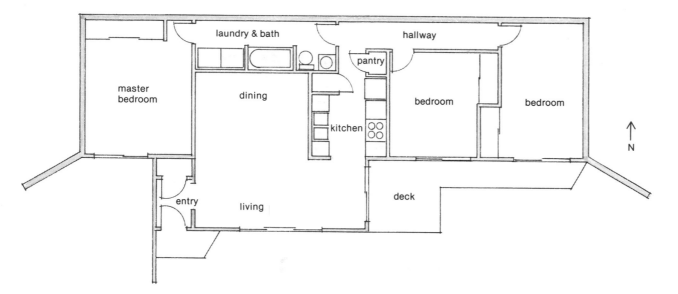

SubT floor plan

concrete floor slab, and fiberglass insulation above the living space equivalent to R 38 at least (7e).

The roof overhang protecting tall patio doors along the south side of the house would vary between thirty and thirty-six inches for most northern latitudes (6g). The first batch of these homes was finished before the winter of 1979.

Besides passive solar input for heating, early versions of the SubT will be electrically heated. But Jeff Sikora predicts that many in-earth solar homes of the future will feature a ceramic-type wood stove in the center of the house with electric back-up heat.

He also predicts that subterranean housing will offer "good deals" to American families of modest income. Guaranteed privacy (2k), energy savings (3h), and the opportunity to utilize otherwise unattractive building sites (2a) are the biggest selling points.

Sikora conceded recently that wood foundations cost just about as much as concrete walls, what with the rapidly inflating costs of lumber (3g). Pressure-treated lumber is estimated to be 20 percent more expensive than regular. Yet, HRB still believes in wood foundations — waterproofed with heavy polyethylene sheets stuck to asphalt tar — as a way to cut finishing costs inside the shell.

SubT is a halfway house. It, and houses like it, will be stopping-off places for families who are refugees of the losing war between overbuildings and cold air. Those of us not quite ready to plunge into a Davis Cave might find SubT a place to get our underground sea legs.

9f. DON METZ

Not long ago, in an address to a "Going Under to Stay on Top" conference at the University of Massachusetts (1e), architect Don Metz said, "The state of the underground art at this point is the result of engineering input. . . . It's functional, but often unimaginative and ugly. . . . We can't forget van Gogh and Beethoven."

Metz's artistry, borne out of his background as both sculptor and builder, joins sound engineering principles, an acute awareness for light and temperature, and sympathy for human convenience. This may be what prompts curious strangers to pound rudely on the door of Baldtop Dugout at 6:50 AM. It's certainly what brought more than 300 people to Lyme, New Hampshire, for a brief open house in Earthtech 5, just after it was completed in August 1979 (1a).

The Winston House (6e), just a few miles from Earthtech 5, was built seven years earlier, in 1972. It's one of the oldest below-grade houses around, but it's timeless in the classic elegance of its elevational design. This, plus its spectacular view, makes it my favorite.

Take Mrs. Winston aside and ask what she'd

Ivy growing up the southwest corner of the Winston House already helps to shade the study window.

The Winston House, Lyme, New Hampshire

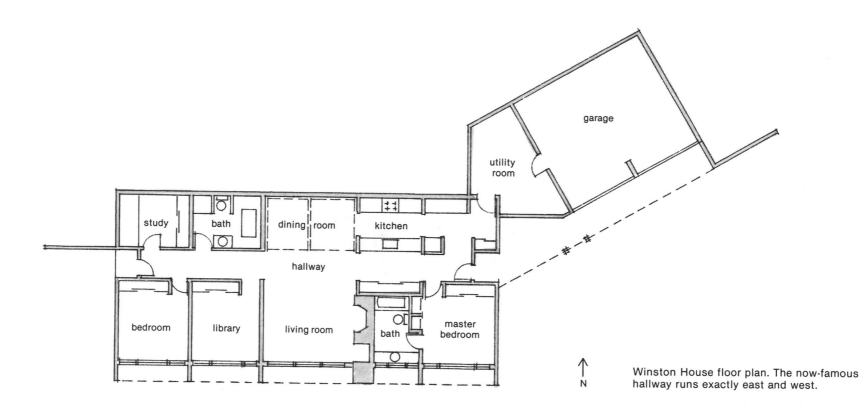

Winston House floor plan. The now-famous hallway runs exactly east and west.

do to improve her house and she'll reply, "Nothing. We trusted our architect who became our friend. This house is perfect for us. . . . Oh, I'd like to change the electric stovetop and the oven. But Don had *me* choose those."

Ask Mr. Winston, an architect himself, if a woodstove might be more efficient than the fireplace in the living room, and he'll say, "No. The house holds heat so well, I don't worry much about possible heat loss through the fireplace. For me an open fire is something to look at and enjoy. As long as I can afford it, I'll

buy firewood for the pleasure of watching it burn."

Each of the Metz houses has things in common: all have concrete walls and wooden roofs supported by rough-sawn 6×10's. They can bear at least 230 lbs/ft^2 of roof load, assuming twelve inches of saturated soil, snow, and any other live loads (6i). All have south-facing windows beneath overhangs, and heat-retaining masonry masses inside (7a).

Portions of the floors are tile, which can collect heat. These tiles are set in epoxy mastic. The sidewall insulation (7e) for the newer

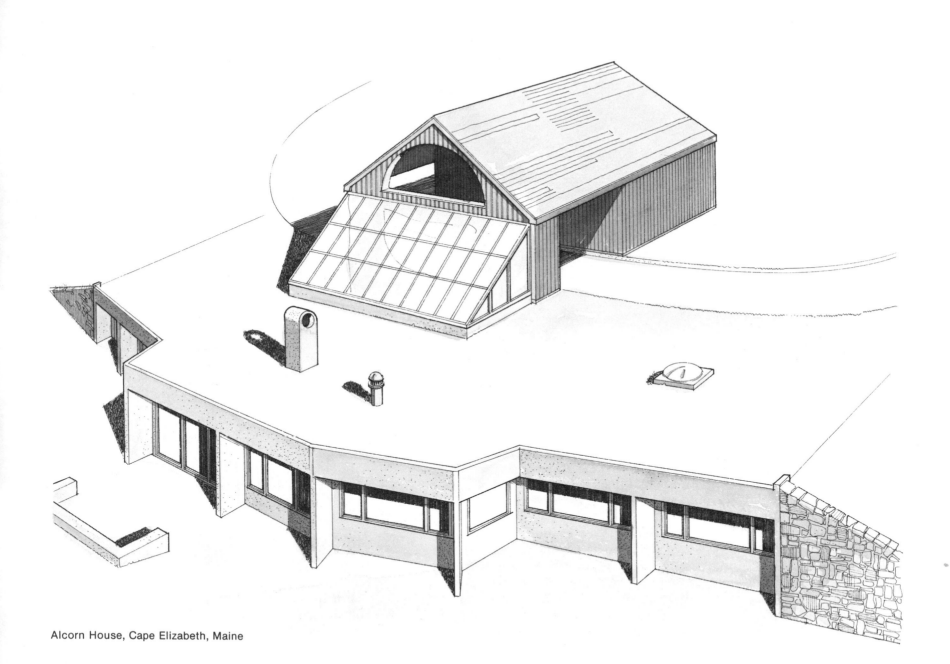

Alcorn House, Cape Elizabeth, Maine

houses is protected on the outside by $^3/_{16}$-inch asbestos board extending two feet or so below the pressure-treated wood sill (6v).

One of the latest Metz designs, the Alcorn House (6f), which overlooks a lovely salt marsh in Cape Elizabeth, Maine, is larger than Baldtop Dugout, the Winston House, or any of the Earthtech homes. Because it shelters the couple and four sons, the house has three bedrooms and a study, two living rooms, two baths, three stoves, two skylights and an exterior wood chute into the mechanical room.

Visitors approach the Alcorn House from above, and are confronted only by the large

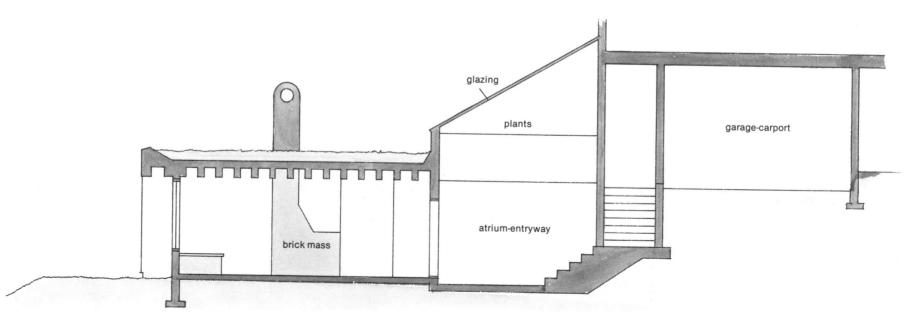

Alcorn House, cross-section showing atrium entryway

two-car garage. Entry is through an immense solar-heated greenhouse complete with growing plants on an upper level and a hot tub in the lower "atrium." The insulated north wall of this solarium is lined with water cannisters painted flat black to provide thermal mass and radiate heat back to this space when the sun isn't shining (7f).

The buried interior is much simpler than it first appears. As in Baldtop Dugout, Metz juxtaposes curving walls and straight walls to influence movement and spice the common spaces with unexpected planes and textures (8c). The Garrison wood stove in the main living room is set against a rectilinear brick hearth and chimney mass. But a guest in the living room can see past this squareness to a round wall — also of Belgian brick and dark mortar — which actually backs an arched brick hood surrounding the Franco-Belge cookstove in the kitchen.

Much of the wood in the house, including the roof beams, is southern yellow pine, a favorite of Bill Alcorn's. The front of the house has three distinctly oriented glazed surfaces between protruding fins — one to receive sunlight in the morning, one at midday, and one in the afternoon.

Contrasting wood stove locations in the Alcorn House

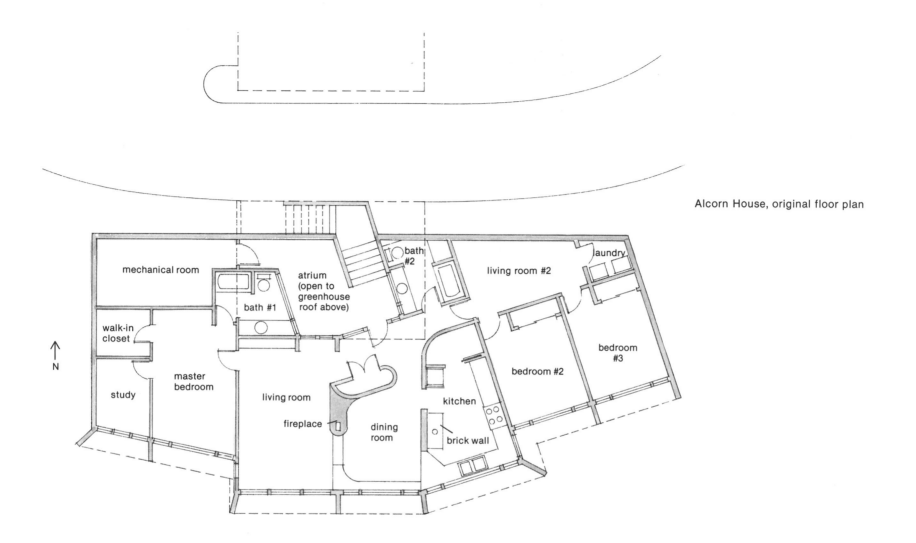

Alcorn House, original floor plan

Labels on plan: mechanical room · atrium (open to greenhouse roof above) · bath #2 · living room #2 · laundry · bath #1 · walk-in closet · master bedroom · study · living room · fireplace · dining room · kitchen · brick wall · bedroom #2 · bedroom #3 · N

This is surely the most lavish — and expensive — underhome I've seen. At first the bank was unreceptive to the idea of their underground house. But after seeing Metz's plans and talking to bankers in Minnesota who financed other earth-sheltered homes (3a), they agreed to the loan. Alcorn functioned as his own general contractor during construction.

Bill and his wife, Susan White-Alcorn, are just beginning to realize they own an instantly famous house. They hope there won't be early-morning intruders.

10 CASE STUDY

10a. CASE STUDY

As a man, I know little about pregnancy. But as research for this book progressed, my gut experienced a growing conviction that I was to build an earth-sheltered home myself. First notice of that seed destined to become Earthtech 6, took place just outside the door of Baldtop Dugout on a chilly spring afternoon. My first meeting with Don Metz, I guess, marked the real end of my journey across continent.

Later, throughout summer's heat, I watched the sun and wandered the high meadows, pastures and forests of central Vermont, looking for a new homeplace — drenching pantlegs in morning-wet ferns, startling partridges from noontime naps, nearly losing my way at twilight. I found several lots I'd be proud to own.

Almost every Realtor in Stowe and surrounding towns was told what I was after, and what it was for: five to ten acres of accessible land, sloping to the south — at a fair price. For an underground house.

Each real estate agent but one — accustomed, it would seem, to large sales and the fat commissions to be made in a resort area —

failed to respond at all to my modest request.

Despite a past winter that was poor for skiing (a mainstay of the local economy), and despite louder and louder signals of impending recession, real estate people, as yet, weren't well versed in the energy benefits offered by passive solar heating within a sunny hillside. I wasn't surprised.

So I bought a cheap pocket compass and searched on my own, following leads from the one sympathetic real estate lady who came up with listings for reasonably priced land. On back roads I talked with locals and farmers who'd put a foot on my car bumper to shoot the breeze about land prices, construction, and the cost of firewood. These people understood better.

By the time yellow school buses again filled the roads at the beginning and end of each day, I felt my long drives and hikes were paying off, and I asked Don Metz to come over from New Hampshire.

Next morning I heard his 750cc BMW purr into the foggy driveway of our rented basement apartment, just seconds before he appeared at the door, removing his helmet, blowing on cold knuckles.

That day, as the fog burned off, we visited four sites I'd selected.

10b. SITES

Sites that were flat, sites that didn't face south, sites that were obviously wet or ledgy, and sites too close to neighbors were eliminated. Those that remained were studied for architectural possibilities — by my untrained eye.

Since I knew the eventual house, although underground, would also have some overground features — a garage and greenhouse, for instance — I was attracted to lots that could be approached from above.

The first site Don and I looked at was off a paved road in Morrisville, the township adjacent to Stowe. It was easy to find our way into the thick of this neglected woodlot because we followed a row of surveyors' "flags" — bits of fluorescent tape hung at eye level in tree branches.

A hundred yards into the woods, there were clearings, and the land crested gently to the south. Here there's room for several well-spaced underhomes along a pleasant wooded knoll. Some clearing would be needed.

Don thought there were very good possibilities here, and I gave him some history:

I knew the land well. In fact, I'd paid to have a five-acre piece surveyed. It was part of a larger twenty-five-acre rectangle belonging to a close friend who'd recently divorced and moved to New Mexico.

Our personal friendship and his need to sell inspired him (as well as his ex-wife) to let me buy the knoll, and a right-of-way to it, at a bargain price. I'd pay for the survey.

"Better check with the Soil Conservation people, though," said my friend over the phone from New Mexico. Clue.

"Right," I said, excited.

But I didn't — until the morning of the survey. Several hours into the job, the surveyor, a forthright man, began to express concern about the soil's ability to percolate, or absorb moisture. There were plenty of trees and other vegetation, but he kept pointing out small, mossy outcroppings of rock, and mentioned that some of the neighbors had trouble meeting Vermont's tough environmental standards for sewage disposal.

Suddenly feeling cold feet, I walked back to the car and drove to the Lamoille County Agricultural Office. Together with the soil conservationist, we looked for the land on the soil survey map.

"Here it is," I said. "What's 13D mean?"

Before our eyes met, he was already shaking his head no.

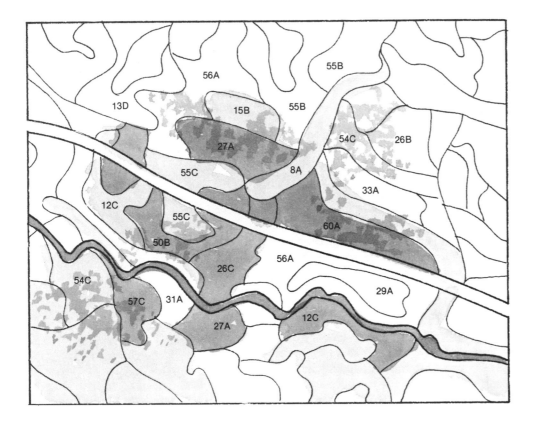

cost of the land. Buying an additional five acres would put the parcel into a more lenient subdivision classification for lots larger than ten acres, but there might always be odor problems because of the thinly masked ledge.

Still, thought Metz, blasting for a house *and* a septic system would not be out of the question, given the otherwise vast potential of the lot.

We went on.

Next stop: a very different site on the opposite side of Stowe village. Here 5.4 acres were for sale. This was woodland, too, though better managed, partly by its present owner, more by the lot's nonhuman inhabitants. Much undergrowth had been gnawed off, then dragged or floated to a stream that passed through the lot, isolating the upper third of the land. Deep within it then, this self-contained plot had a series of reflecting pools, fashioned flawlessly by beavers.

The soil here was adequately rich, and deep enough, according to my new acquaintance, the soil conservationist. In fact, the lot had already been approved for septic tank use.

A long and rather costly driveway might dip into this land from the secondary road where Don and I parked. With a wide sweep to the

"Bedrock," he said. "That's basically the code for too little earth cover. The deepest deposits have twenty to forty inches of soil. Not deep enough for a septic system. To bury a septic tank takes at least four feet. Usually more. Sewage would have no place to go."

Consultation with an engineer friend and much on-site poking with a crowbar confirmed the Department of Agriculture's survey. Engineering an acceptable septic system would be complicated, more than offsetting the low north, this drive could cross an earthen bridge incorporated into the lowest beaver dam, skirt the pond without disturbing it much, to approach the hillside on the far shore.

A small but protected homestead might have looked inward on this sunny glade. There would be no view to the outside, but the rocks, trees and stream would provide enough visual reward. Plus, there would be little wind. Yet somehow we found ourselves lowering our voices to discuss the intrusion of a house in this ready-made Japanese garden.

Oddly enough, it was the site's most attractive ingredient, open water, that caused its rejection. To keep an underground home above water table and dry, it would have to be located up the hill from the ponds — to a point that crowded the property line. Lower on the slope might be an ideal spot for an above-grade solar home, but this was not the best place for earth-sheltering.

As we walked out, Don and I agreed to leave the beavers unbothered — at least by us.

The third site I wanted Don to see was discovered by accident, at a time I was discouraged. I was driving through a fringe of Stowe to see the engineer who was to tell me

the lot I'd just paid to survey was no good. A red flash in the ditch caught a corner of my eye.

I backed up, got out, and poked through the weeds. Sure enough: a FOR SALE sign, obscured by the tall grass. I clambered up the bank and peered into the woods. A southern exposure, all right, but too steep and too ledgy. I'd already learned my lesson about ledge. But I wrote down the telephone number, noticing the exchange of a nearby town written in Magic Marker on the sign.

A week or so later I called. The lively Vermont gentleman on the other end of the line said he was seventy-five, and that he'd be pleased to sell me all fifty-eight acres of that land, just to get it off his hands.

I allowed as how I wasn't interested in that much acreage, and that what I'd seen didn't look right for what I wanted anyway. But somehow he talked me into meeting him to walk the land.

He showed up in a large Mercedes. We spread some drawings of Earthtech 5 on the hood of his car. He was interested.

"Let's have a look at this hill," he said, grabbing his walking stick, and striking out up the bank. All I could do was plod along after him.

A quarter of a mile up the rocky slope, the

terrain changed unexpectedly, and flattened just as we broke from the trees onto softly rolling meadowland. Beautiful. Without breaking stride, he led me across the huge clearing to a sloping spot with a spectacular 360-degree view. Mountains and villages in all directions. It seemed a perfect site, and my compass agreed. So did the soil maps I studied later that afternoon.

I tried not to seem too impressed.

The old man watched me closely as he showed me two other sites, each as magnificent. He sat quietly on a rock while I explored more closely on my own. Then he took me down off his hill and bought me lunch. And a week or so later I took him to Lyme, New Hampshire, to meet Don Metz. That day I bought *him* lunch, and asked if he'd sell me a small part of that land. For an underground house. He was interested, he said.

What this splendid old businessman *really* wanted was to sell me the entire parcel — one of his many real estate holdings in Vermont. But of course, it was hopelessly beyond my financial reach. When I called him weeks later to ask permission to walk the land with Don Metz, he said the entire fifty-eight acres was in the process of being sold. But he had the owner-to-be contact me.

When Don and I looked at this land together, it was with the young man who plans to develop it. He, too, was interested, but an access road up the steep slope to the high meadow is an expensive proposition. So is bringing in electrical service. This new owner, who wants to do things right, was not ready to commit to a price for a five-acre lot. Besides, he was thinking about building a house near the same spot I would.

The right site, ruled out by poor timing and wrong circumstances.

About half a mile up the road from the beaver-pond place in Stowe Hollow, there was another 5.28 acres for sale. The price seemed too high. But I kept going back there on my way to work in the morning; in the afternoon on my way home; in the middle of a rainy day. Another day I circled it again and again in a small rented Cessna. Deep inside, I knew it was the prettiest of all.

Approached from the town road above, a workroad beckons you onto the lot. The grass strip between two tire marks overcomes them eventually, making the road disappear beside a grove of immense trees, spared by generations of woodcutters. These ancient sugar maples, each at least four feet in diameter, form an unkempt

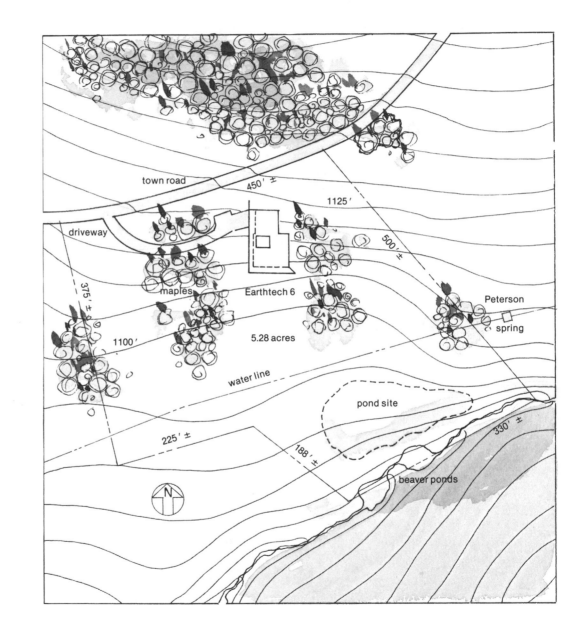

Campbell site plan

corridor framing a peak in the distance. The mountain, known as The Pinnacle, is dead south.

The soil survey map finds this land to contain "Berkshire Stony," a soil type described this way by the U.S. Department of Agriculture:

> Soils in this series are well drained, loamy and stony. . . . These soils occupy glacial till covered uplands with slopes of 0 to more than 65 percent. These soils typically have fine sandy loam or loam surface layers and subsoil that overlie fine sandy loam at a depth from about 22 to 50 inches. Cobbles and stones are present throughout the profile. Permeability is moderate to moderately rapid; available moisture capacity is medium and natural fertility is low. Depth to water table and bedrock typically exceeds 5 feet.

Don Metz stood with his hands in his pockets and whistled softly. "Very dram-a-tic," he muttered. Impressed.

Electrical lines are close by. The work road can be easily improved. According to the neighbors, the sun never drops below the Pinnacle, even at its lowest altitude just before Christmas. The slope is right, and the maple trees west of the house will lose their foliage each fall, to make way for solar radiation passing through on its way to a glazed west wall.

Campbell site

Sun will reach the house from the south at all times of year.

The soil isn't the most fertile, but that's easy to correct. And the brook that feeds that marshy corner of the lot runs strong, even through a dry August. I've checked. When the excavation is done, a pond can be dredged at the same time and the dug-out earth used to bury the house later on . . .

Metz said it was time to go home to his drawing board. He'd seen what he'd come for.

Earthtech 6 will be here.

10c. EARTHTECH 6

Our home must ease softly into this lot — so the deer, game birds, beavers, foxes, and other creatures who already live there, hardly take notice. From the ground it will be inconspicuous, and from the air, all but impossible to spot.

The basic envelope of the house will resemble Earthtech 5, tailored to suit both Carol, the lady who shares my home as well as my time, and me. Though small, it must also serve the needs of my two children (frequent residents), a steady stream of house guests, and, however painful it

Campbell House, exposed west elevation

may be to think about them, some unforeseeable owners in the future.

Before he drew anything, Don Metz needed to learn all these things. And much more, of course. Designing Earthtech 6 together has made us friends.

The improved driveway will be fairly flat as it finds its way to a double garage door set in the hillside beneath its overhang draped with ivy. The garage entrance and the front doorway to its right both face west.

The garage and the house itself, nestled somewhat lower in the slope, will be earth covered, with roofs essentially flat. Connecting these two major elements of the structure will be what Metz has dubbed a "light funnel,"

shaped like a triangle when seen from the driveway.

The roof of this long directional skylight pitches at twenty-two degrees — the sun's exact altitude in Stowe, Vermont on the winter solstice, December 21. This small shed roof will be shingled while the remaining roofs will be disguised in vegetation — grass, ferns, and small shrubs to match the rest of the lot.

The land drains well, but special precautions must be taken to berm and swale the buried north and east sides of the house to flow surface water away from the living space. Perimeter drains, encircling the footings, will be covered with sand and gravel above to insure that underground water is lured away from the walls.

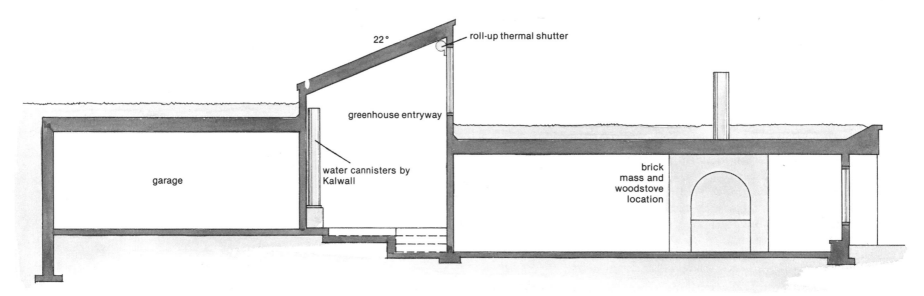

22°

roll-up thermal shutter

greenhouse entryway

water cannisters by
Kalwall

garage

brick
mass and
woodstove
location

Campbell House, cross-section

These walls, reinforced cast-in-place concrete, will be waterproofed with Bentonite. The roofs, made of hemlock timbers and planks in typical Metz fashion, will be sealed with an impervious bituthane membrane. Roof water will run into drains in interior walls.

A visitor outside will immediately recognize the open south and west walls. Each is filled with large windows, protected by wind fins, veiled by overhang in summer heat, coyly inviting winter sunlight.

Retaining walls at the northwest, southwest, and southeast corners of the building will resist earthflow from above, while a shorter wingwall, made of stone, also identifies the transition between floor levels inside.

All exterior surfaces not glass or concrete will wear cedar shiplap siding, installed vertically, then bleached light grey and left to weather in whatever way Mother Nature dictates. One flat skylight will sit above the kitchen, neighbor to one chimney and several stacks that also penetrate the roof.

Those who come into the building by car, on one hand, can enter the lower level through a door at the back of the garage, walk down a set of steps and through the mechanical-storage room. Guests, on the other hand, might park outside and arrive through the greenhouse. Once inside, they'll pass down angled steps into the angled vestibule at the rear of the living room.

Campbell House, floor plan

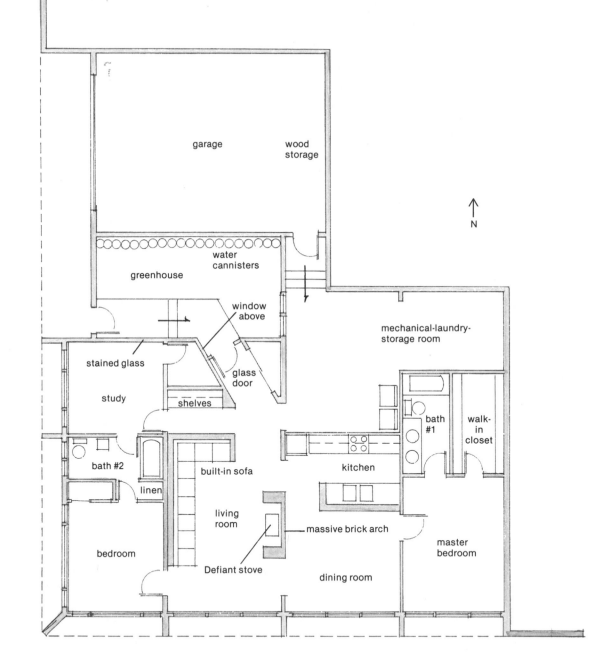

garage

wood storage

water cannisters

greenhouse

window above

mechanical-laundry-storage room

stained glass

glass door

study

shelves

bath #1

walk-in closet

bath #2

built-in sofa

kitchen

linen

living room

master bedroom

bedroom

massive brick arch

Defiant stove

dining room

N

The insulated north wall of the greenhouse is to be lined with water-filled cylinders from Kalwall, a manufacturer of solar age plastic products in Manchester, New Hampshire. These will be painted black to absorb radiation and hold heat from the sun. Basically, the water cannisters will serve as increased thermal mass.

To save heat and reduce exposed surface area, the windows on the south wall of this solarium will be standard double-glazed windows, mounted vertically. In summer, they can be opened to pull air from the lower house, and offer cross-ventilation as heat rises out of the light funnel.

It's our wish that this greenhouse, beyond lending light and heat to the back of the house, will allow us to grow a few varieties of vegetables year round — as well as ornamentals. The mechanical room will be lighted from the greenhouse through a frosted window, as will the study-bedroom, through stained glass.

In winter, direct sunlight will reach deep into the master bedroom, dining room, living room, and guest bedroom through the south and west walls. Here, heat from solar radiation will be stored in concrete, in floor slate, massive roof timbers and in the large masonry arch over the central woodstove.

Back-up heat will come from electric

Campbell House, Stowe, Vermont

baseboard units, metered separately from the rest of the appliances. We hope it will almost never be needed — at least while the woodstove is burning or the sun is out. The house will be insulated outside the concrete with rigid, closed-cell synthetic foam sheets.

Walls will be painted off-white to enhance the already bright interior. Floors will have carpet, tile, and hardwood. Some furniture in the living room and work surfaces in the bedroom study will be built in, designed by Metz to divide space comfortably and still be in keeping with the clean lines characterizing the rest of the house.

We're anxious to start construction on this earth-protected home in the spring. In our heads, at least, we've already settled in there.

APPENDIX

APPENDIX 1

Where to Insulate Earth-Protected Buildings and Existing Basements

THOMAS P. BLIGH [1], PAUL SHIPP [2] AND GEORGE MEIXEL [3]

INTRODUCTION

The extraordinary interest in earth sheltered housing is due primarily to its substantial energy savings compared to even the best conventional structures. In typical above-ground construction, energy is wasted by unwanted heating and cooling of the surroundings. By reducing heat transfer to and from the surroundings, less energy is required to maintain desired conditions. Heat loss (or gain) from a structure

1. Department of Mechanical Engineering, MIT, Cambridge, MA 02139.

2. Department of Mechanical Engineering, University of Minnesota, Minneapolis, MN 55455.

3. Department of Civil and Mineral Engineering, University of Minnesota, Minneapolis, MN 55455.

principally depends on two factors: the ventilation load for heating or cooling intake air, and the heat transmission through the building envelope. In most residences, the ventilation load consists of uncontrolled air infiltration through cracks and holes. These unwanted infiltration losses are practically eliminated by earth covered construction. Ventilation air can then be controlled so that heat recovery systems can be used effectively to transport heat from the exhaust to the inlet air. The transmission losses depend on the amount of heat conducted through the envelope of the structure. This is a function of the thermal transmission coefficient, which is reduced by adding insulation, and the temperature difference between the inside and outside of the wall.

Above-ground, temperature difference is determined by the local weather conditions. On the other hand, the earth averages the temperature fluctuations both on a daily and yearly basis. Seasonal temperature fluctuations reach a depth of several meters into the soil, whereas the penetration of short-term temperature fluctuations over periods of hours or days is almost negligible. At depths of 3 meters or more, the temperature fluctuates only slightly either side of a constant value a few degrees above the yearly average (for example, $10°$ to $12°C$ for the northern states). For a detailed discussion, see Reference 1.

When the wind blows against an above-ground house, a high pressure zone is formed on the windward side forcing cold air through all the inevitable cracks, while a low pressure zone is formed on the leeward side which sucks warmed air out of the house. In addition to the increased infiltration, the wind continually blows away the surface layer of air which has been warmed by conduction through the wall. This increases the surface heat transfer coefficient and more heat is lost by conduction due to this "wind-chill" factor. Earth sheltered houses have far fewer exposed areas so these effects are reduced dramatically.

Finally, an important factor in controlling the energy required to heat or cool a building is the "thermal mass," or total heat capacity, involved. A building having a large thermal mass within the insulation can store a large amount of energy so that during the day energy entering the south solar passive windows can be stored, and the temperature will rise slightly and slowly throughout the day. During the night there will be a net heat loss from the building envelope and heat will be transferred from the thermal mass to the inside air and the temperature will fall slowly throughout the night. On the other hand, a building with a low thermal mass cannot store much energy per degree rise, and therefore the incoming solar energy will quickly heat the building to an uncomfortably high temperature, possibly leading to a local cooling requirement. Since little energy is stored, the temperature will fall rapidly during the night and more heating will be required.

The soil surrounding a building constitutes a large thermal mass which reduces the temperature variation felt by the exterior structure and can be used for passive solar energy storage.

In addition to these energy benefits, the earth protects the building from expansion and contraction, and especially from freeze/thaw damage. It is also worth emphasizing that in the case of a power failure during extremely cold weather, the temperature within an earth protected building will still be over 10°C (50°F) even over long periods. Surface buildings under these conditions rapidly drop in temperature, pipes freeze causing great damage, and the building is not habitable within hours. Earth protection therefore reduces dependence on an external energy source from a matter of survival to one of comfort control.

ENERGY COMPARISON

The energy-saving advantages of earth sheltered designs can be illustrated by comparing the total winter and summer energy use of single and two level houses built either above or below ground. The detailed assumptions and calculation method are given in Reference 2. The values used were for Minneapolis, Minnesota, 44° North latitude. The heat gain/loss through the windows, walls and roof for the above ground houses and the south-facing wall of the below ground house were calculated using data from ASHRAE, Reference 3. The south side double-glazed windows were 35% of the south wall area, and drapes were drawn from dawn to dusk which reduced the daytime U-value of 3.349 W/m^2 K (0.59 BTU/ft^2 °F) to a nighttime U-value of 2.555 W/m^2 K (0.45 BTU/ft^2 °F). The windows had a shade factor of 0.83. The solar intensities used were those given in ASHRAE, Table 2.

It is worth noting that the seasonal load values are based on the average monthly performance, and therefore these values are valid estimates of the seasonal energy cost for each structure, and are a good basis for comparison. The peak loads for extreme weather conditions were lost in the averaging process, so these results do not represent design values for sizing HVAC systems. If real extreme temperature fluctuations had been included, the earth sheltered structures would compare even more favorably, since daily temperature changes have little impact on the high mass structures.

The soil markedly reduces the amplitude of any air temperature swing, and produces a phase lag so that the peak conduction loss in an underground building does not occur at the same time as the peak load due to the cold infiltration and ventilation air. Clearly in an above ground building, the peak conduction and air-tempering loads coincide, and therefore a larger HVAC system must be used, and this will then cycle on and off frequently which reduces its overall efficiency

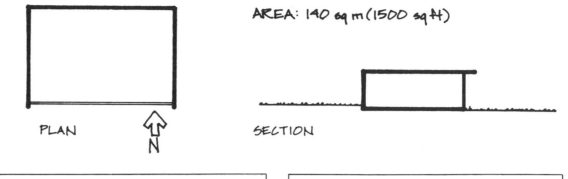

AREA: 140 sq m (1500 sq ft)

PLAN ⇑ N SECTION

Figure 1. Energy use of an above-ground, one-level, slab-on-grade house. Windows: 35% of south wall area.

B-1. Walls: 2×4 in lumber with 3½ in fiberglass.
Roof: 2×6 in lumber with 5½ in fiberglass.

B-2. Walls: 2×6 in lumber with 5½ in fiberglass.
Roof: 2×10 in lumber with 9 in fiberglass.

B-1	WINTER (kW-hr)	SUMMER (kW-hr)
transmission	−12935	+3658
ventilation	− 1544	−
internal heat	+ 5900	+1700
net energy	− 8579	+5358

B-2	WINTER (kW-hr)	SUMMER (kW-hr)
transmission	−10186	+3173
ventilation	− 1544	−
internal heat	+ 5900	+1700
net energy	− 5830	+4873

substantially. The increased operating efficiency of underground HVAC systems was not included in the tables in which a negative sign indicates heat loss.

The first example compares a one level 140 m² (1500 ft²), conventional above ground, slab-on-grade house (Figure 1) to a similar structure which was earth sheltered, shown in Figure 2. The earth sheltered structure has earth placed against the north, east, and west walls and the roof consists of 50 cm (20 in) of soil placed over 10 cm (4 in) of polystyrene insulation on a 20 cm (8 in) precast concrete plank. The same insulation is placed against the top 155 cm (5 ft) of wall before back filling. During the seven winter months, the inside temperature was held at 20°C, and during summer at 25.6°C.

Note that the transmission losses in the earth sheltered house are so small during winter that the heat gained from the internal load and the passive solar windows is greater, on average, than the heat loss through the building envelope. During cloudy weather and extremely cold snaps, a small backup

Figure 2. Energy use of an earth-sheltered, one-level house.
Windows: 35% of south wall area.
Walls: Soil, 10 cm polystyrene for 155 cm down from wall top.
Roof: 50 cm soil, 10 cm polystyrene, 20 cm concrete plank.

	WINTER (kW-hr)	SUMMER (kW-hr)
transmission	−4043	− 969
ventilation	−1544	−
internal heat	+5900	+1700
net energy	+ 313	+ 731

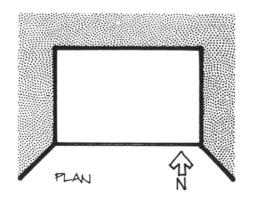

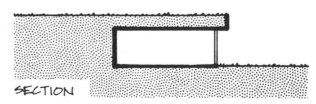

AREA: 140 sq m (1500 sq ft)

PLAN ⇑ N SECTION

Figure 3. Energy use of an above-ground house with two levels and walkout basement. Windows: 35% of south wall area, both levels.

F1. Walls: 2×4 in lumber with 3½ in fiberglass.
Roof: 2×6 in lumber with 5½ in fiberglass.

F2. Walls: 2×6 in lumber with 5½ in fiberglass.
Roof: 2×10 in lumber with 9 in fiberglass.

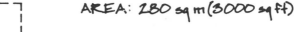

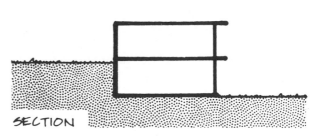

AREA: 280 sq m (3000 sq ft)

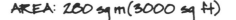

F-1	WINTER (kW-hr)	SUMMER (kW-hr)
transmission	−9876	+3358
ventilation	−3087	—
internal heat	+6600	+1800
net energy	−6363	+5158

F-2	WINTER (kW-hr)	SUMMER (kW-hr)
transmission	−7136	+2872
ventilation	−3087	—
internal heat	+6600	+1800
net energy	−3623	+4672

heating system, such as a wood burning stove or small conventional furnace will be required. During the same seven winter months, the above ground houses have large net energy deficits and will require a large furnace to meet the high peak loads during cold snaps. Even the well-insulated house, Case 2, has a substantial heating and cooling load. In fact, above ground houses almost invariably have windows or doors on the east, west, and north sides. This leads to an increase in the transmission and infiltration losses and thus real above ground houses have loads even higher than those shown in Figures 1 and 3.

During summer, the surrounding cool soil continues to cool the earth sheltered house at a rate slightly less than during winter. Above ground houses now are heated by the sun and infiltrated air which, added to the internal load, produces heat gains six to eight times greater than that for the earth sheltered designs.

Figure 3 shows a conventional above

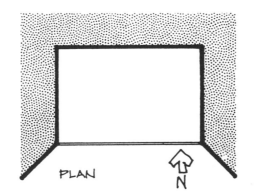

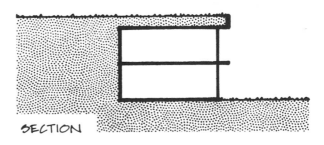

AREA: 280 sq m (3000 sq ft)

Figure 4. Energy use of an earth-sheltered two-level house.
Windows: 35% of south wall area, both levels.
Walls: Soil, 10 cm polystyrene for 155 cm down from wall top.
Roof: 50 cm soil, 10 cm polystyrene, 20 cm concrete plank.

	WINTER (kW-hr)	SUMMER (kW-hr)
transmission	−3884	− 634
ventilation	−3087	—
internal heat	+6600	+1800
net energy	− 371	+1166

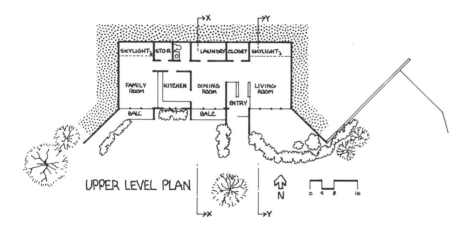

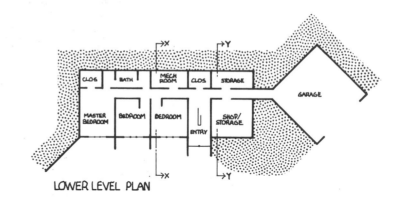

Figure 5. Upper and lower level plans of a two-level, earth-sheltered house.

ground house with a walkout basement, a two level design frequently used in the northern climates. The table shows that the below ground section considerably improves the performance over a slab-on-grade structure. The above ground portion so handicaps the building, however, that it falls far short of the performance of the earth sheltered two level example shown in Figure 4. This illustrates that the plan area of the house shown in Figure 2 can be doubled without incurring a real energy penalty. The remarkably low energy use is due to the more compact configuration than the single level and much of the envelope is deeper and better protected from temperature swings. The two level contains twice the floor area but has only 30% more total surface area through which heat can be transmitted. Thus despite increased ventilation and internal heat loads, the building still has a very low winter and summer energy consumption.

An important issue in areas subjected to

blizzard conditions is the rate at which the inside temperature will drop in the event of a power failure. Taking the maximum rate of heat loss, in January, for the house shown in Figure 2, and providing no allowance for any heat gain, such as from people or a wood burning stove, the temperature of the building will drop at a rate of less than $1°C$ per day. Not only is safety provided, but habitable conditions could be maintained almost indefinitely with, for example, a small wood burning stove.

In all of the above studies a sandy loam soil was chosen with a thermal conductivity of 0.81 W/m K (0.47 BTU/hr ft°F). The specific heat and density were 858 J/kg K (0.205 BTU/lb °F) and 185 kg/m³ (113 lb/ft³). As moisture migration in soils is complex and not well understood, a constant value was used. One design was analyzed with moisture contents ranging from 18% to 37% and it was found that the winter heat loss increased by 28%, while the summer

cooling rate increased by 49% for the wet soil. This does not have a large impact since the transmission losses are small for earth protected structures. However, it should be noted that in northern climates the soil will be wet in summer from snow melt and rain, and the heat loss will be higher, to advantage for summer cooling. During the winter, the surface freezes and the temperature gradient slowly drives the moisture away from the building, drying out the soil; the heat loss is then reduced, again to advantage. The thermal conductivity of soils can change by as much as a factor of 10 with increasing moisture content, so it is important to know the moisture content with reasonable accuracy. Data correlating the thermal conductivity with moisture content for various soils is given in Reference 4.

In Figures 5 and 6, Carmody and Ellison's design of a two level earth sheltered house is given to illustrate a real design similar to that analyzed in Figure 4.

Figure 6. General view and sections of a two-level, earth-sheltered house.

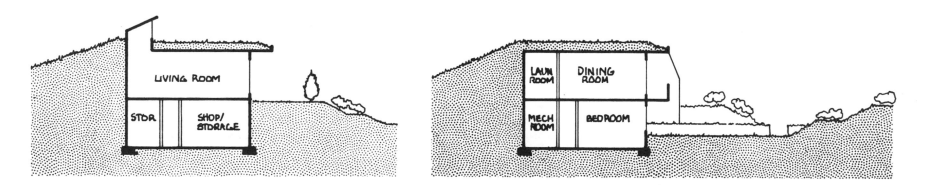

LIVING ROOM

STOR. SHOP/ STORAGE

LAUN ROOM DINING ROOM

MECH ROOM BEDROOM

ROOF INSULATION

Because of the small transmission losses through the walls and floor of an earth sheltered structure, the roof heat loss may exceed 50% of the total transmission losses. The thermal design of the roof is therefore important. Since the thermal conductivity of soil is about 25 times greater than, say, polystyrene, its insulating properties are not a significant factor in roof design. How much soil and how much insulation then should be used?

The heat transfer performance of the roof can be greatly influenced by two important properties: the heat capacity or thermal mass over the roof, and the surface boundary conditions at the roof/air interface.

The heat flux through various roof cross-sections was examined using a transient, one-dimensional, finite-difference computer program described in Reference 2. The inside air temperature was allowed to vary seasonally and ranged from 20°C during the winter to a peak of 25.6°C in summer. The outside air temperature was simulated by a sinusoidal approximation of the average Minneapolis temperatures from 1940 to 1970. As justified later, no solar gain was included since the surface was covered by grass.

Two roof structures of similar R-values were compared. The first, Roof C, had 3.0 m (9.8 ft) of soil supported by a 31 cm (12 in) precast concrete plank giving an R-value of 4.32 m² K/W (24.51 hr ft² °F/BTU). The second, Roof A, had 46 cm (18 in) of soil over 10 cm (3.9 in) of polystyrene supported by a 20 cm (8 in) concrete plank, giving an R-value of 4.35 m² K/W (24.68 hr ft² °F/BTU). Over the seven winter months, the 3.0 m soil roof showed a mere 2.4% reduction in transmission loss. A soil depth of 46 cm is normally required to sustain good growth in most areas and this load can be supported over reasonably large spans (see Reference 2) by 20 cm (8 in) concrete planks. However, it should be noted that with the 3.0 m soil cover, the increased depth of the building would reduce the heat losses through the walls slightly, and the summer cooling effect through the roof would be better. These are fairly small effects which in general would not offset the added cost of the massive structure required.

The above discussion is for the energy losses over a period in which the temperature rises and falls as simulated by the sinusoidal mean and is therefore valid on a monthly or yearly basis. Daily temperatures, however, vary in random intervals of days above or below this mean. To test the effectiveness of the earth cover a cold front was simulated for five days. A mid-January day was selected in which the air temperature varied sinusoidally from an early morning low of −15°C (5°F) to an afternoon high of −9.4°C (15°F). On January 15, a cold front caused a drop of 5.6°C (10°F) for all daily temperatures which then varied sinusoidally in the range −20.6°C to −15.0°C (−5° to 5°F) for five days after which they returned to the normal mean sinusoidal cycle. Figure 7 shows the time responses to these conditions of two roof structures with identical R-values of 4.35 m² K/W; Roof A, and Roof B consisting of 12 cm of polystyrene supported by a 20 cm concrete plank. Note that the figure shows the mean daily temperature.

The daily heat loss per unit area is plotted, hence the shaded areas beneath each line represent the excess heat loss due to the cold front. Due to its low thermal mass, Roof B responds rapidly to the outside change and maintains the new maximum heat loss until the quick return to normal after the cold front passes. The high thermal mass structure A, however, requires a full day before the ceiling begins to indicate that more severe conditions now exist outside and the response is slow. After five days, when outside temperatures return to normal, the heat loss still gradually rises to a maximum of 85% that of Roof B. Roof A then slowly returns to normal.

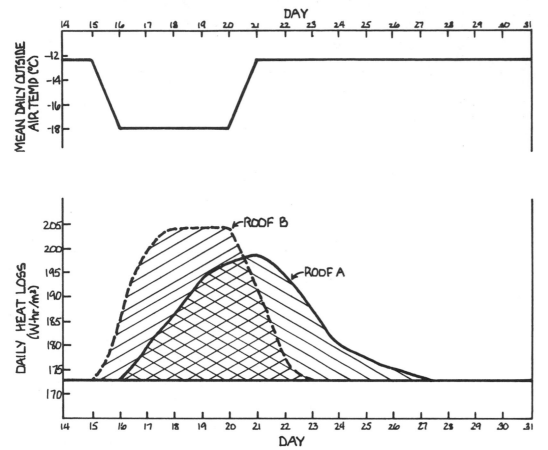

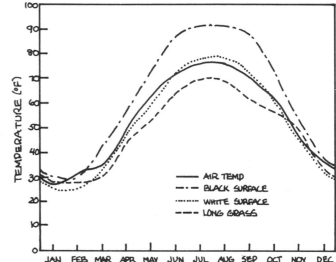

Figure 7. Thermal mass in roof structures: the effect on daily heat loss due to soil cover on a roof subjected to a cold front for five days.

Figure 8. Monthly average temperatures beneath four different surfaces, from Reference 2.

Despite the longer total response time, Roof A required 8% less total energy than B, and the peak load was 15% lower. Furthermore, Roof B's low thermal mass required almost twice as much additional energy (196%) during the five day cold period when ventilation and infiltration air was at its coldest. This does not affect the net energy balance, but does increase the peak load which means a larger heating system must be installed as discussed earlier. This

demonstrates that while two structures may be thermally equivalent under steady conditions, the high mass roof uses less energy and reduces the peak load during short term transient cold or hot snaps.

Finally, vegetation on the roof contributes to the thermal efficiency by shading effects, improved insulation due to air pockets in the foliage and, most important, the elimination of solar heat gain to the roof during summer. This in part is due to reflection and

photosynthesis, but mainly to transpiration which cools the foliage by the latent heat change as moisture evaporates. Temperature measurements beneath paved and grass-covered patches showed that daily high temperatures in summer produced by solar radiant heating can exceed 60°C (140°F) below an asphalt surface even though the air temperature was no more than 32°C (90°F). Under the same conditions the highest temperature beneath grass was only 40°C

Figure 9. Vertical wall insulation comparison.

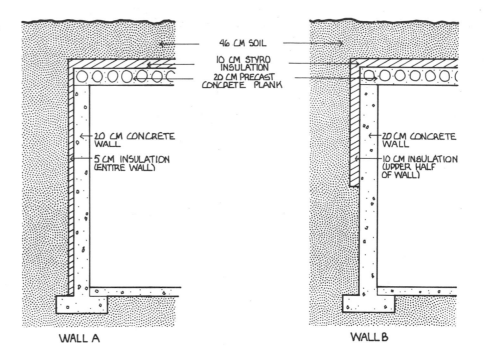

(104°F), see Reference 5. During the summer months the net effect of radiant input to various surfaces is shown in Figure 8. A blacktop covered surface becomes 8.3°C (15°F) warmer than the average air temperature, while grass cover is consistently below ambient conditions by 0.6 to 3.9°C (1 to 7°F), depending on the length of grass. During the winter the grass will provide extra insulating air pockets.

Many factories, warehouses and shopping centers are constructed with flat blacktop roofs. It would be worth while to examine the economic feasibility of converting these to sod covered, energy efficient, low maintenance roofs.

WALL INSULATION

Windows and doors should be restricted as far as possible to the south wall. In winter, double-glazed north-facing windows have an average heat loss flux twenty times that of an earth protected wall; during the coldest weather this can reach a factor of 35. An east or west facing window represents ten times the wall flux. On the other hand, south facing windows can produce a net heat gain, as discussed later.

The soil surrounding an earth protected building will be heated, after some time, to

some temperature between the building temperature and that of the original soil. During the summer the inside temperature of the building should be allowed to reach 26°C. The heat flux into the soil helps to keep the building cool. In winter the inside temperature should be kept at no more than 20°C. For a short while, especially in two level designs, heat will actually flow back from the soil into the building as has been measured in Williamson Hall (Reference 2). For the rest of winter the heat loss into the soil will be less than if the soil had not been preheated the previous summer.

If the walls are totally insulated, the building will be isolated from the soil and will not interact with the soil thermal mass as described above. The question then arises

as to how much soil is required and where it should be placed.

The heat flux into an uninsulated, earth protected wall would be maximum along the top and decrease to a minimum at floor level. Therefore it is most important to increase the thermal resistance near the top of the wall. This can be done by placing insulation vertically down the wall or extending the roof insulation horizontally out from the wall to cut the heat flux paths.

To analyze this, a sample structure with identical amounts of insulation, Figure 9, was used. After three years, to achieve steady operating conditions, wall B showed a 10% improvement in summer cooling at the cost of a 5% increase in winter heat loss. Averaged over the year this came to a

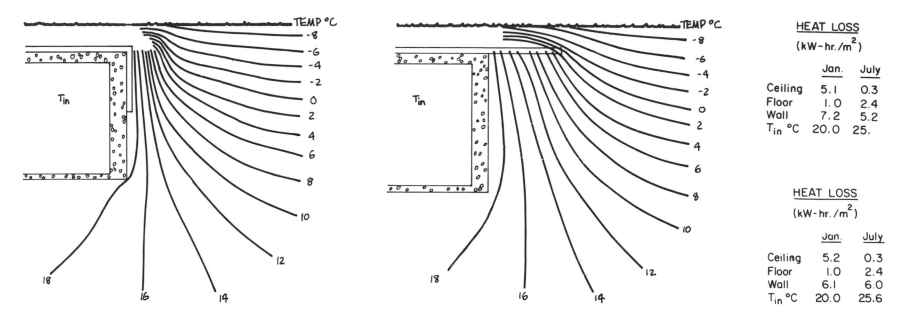

HEAT LOSS		
(kW-hr./m²)		
	Jan.	July
Ceiling	5.1	0.3
Floor	1.0	2.4
Wall	7.2	5.2
T_{in} °C	20.0	25.

HEAT LOSS		
(kW-hr./m²)		
	Jan.	July
Ceiling	5.2	0.3
Floor	1.0	2.4
Wall	6.1	6.0
T_{in} °C	20.0	25.6

Figure 10. Horizontal wall insulation comparison. Isotherms shown in °C. Note that heat flux lines are orthogonal to the isotherms and the horizontal insulation produces a small and uniform heat flux from the wall.

savings of 100 kW-hr during the summer at the expense of 107 kW-hr during the winter for the 140 m² house of Figure 2. While there is no net energy gain, in wall B the advantage has been shifted to summer cooling. If the cooling load can be kept sufficiently low, it will not be necessary to install an air conditioner. The HVAC system then is reduced to a simple heating unit with clear advantages.

Finally the same amount of insulation could be placed horizontally out from the wall as illustrated on the right side of Figure 10. Notice that the heat fluxes through the ceiling and floor are practically identical in the two cases. However, the winter loss through the wall with horizontal insulation

has decreased from 7.2 to 6.0 kW hr/m², an improvement of 20%, and in addition the summer cooling heat loss improved by 17%.

In practice, support must be provided for the insulation at the corner using plywood or flashing. An added advantage of this design is that water will be drained well away from the building wall so that waterproofing the wall will be less critical.

FLOOR AND BASEMENT INSULATION

The heat loss through the floor of an earth sheltered structure will be small and steady year round. Computer studies (Reference 2) show that the rate of loss through an

uninsulated concrete floor is about 1.2 W/m² (0.38 BTU/hr ft²) which represents less than 14% of the total winter's energy loss through the building envelope. The addition of 2.5 cm (1 in) of polystyrene beneath the floor decreases this loss by less than 5%, which is less than a 1% decrease in the total building energy use.

The same amount of insulation added to a roof with an R-value of 4.23 m² K/W results in a 20% decrease in ceiling loss during the winter, an improvement of 11% in overall performance. Thus, it is far more cost effective to add insulation to the roof than to the floor.

This conclusion is reinforced in summer since the thermal mass of the soil below an

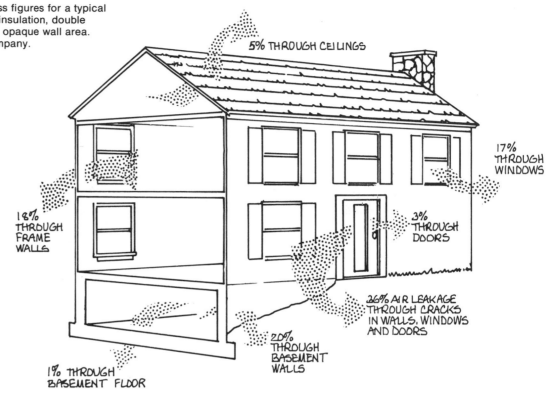

Figure 11. Comparative heat loss figures for a typical two-story house with moderate insulation, double glazing and 186 m² (2000 ft²) of opaque wall area. Source: The Dow Chemical Company.

5% THROUGH CEILINGS

17% THROUGH WINDOWS

18% THROUGH FRAME WALLS

3% THROUGH DOORS

26% AIR LEAKAGE THROUGH CRACKS IN WALLS, WINDOWS AND DOORS

20% THROUGH BASEMENT WALLS

1% THROUGH BASEMENT FLOOR

uninsulated floor allows heat to flow out of the building at a greater rate when the inside temperature increases. Thus a threefold increase in heat loss through the floor can be encountered during the summer. Installing insulation beneath the floor slab significantly reduces this cooling effect.

To alleviate the cold sensation, under floor insulation has been advocated to allow the floor slabs to float more nearly at room temperature. A floor at room temperature, 23°C (73°F) is still some 14°C (26°F) cooler than body temperature. A better solution to the comfort problem is not to insulate the slab as discussed above but to install carpeting, wood flooring or area rugs over the slab. Thus the desired comfort is assured without any appreciable degradation of the positive effects of the floor's thermal behavior.

The above ground house shown in Figure 11 illustrates a number of interesting points pertinent to this discussion. First, the heat loss through the basement floor is essentially negligible for the winter heating load. Two major heat loss areas are the basement walls, mostly near the top, and through air leakage through cracks in the walls. A critical area for these cracks is along the joints between the basement wall and the above ground timber wall. An effective and economical means of dealing with these losses would be

to place 5 cm (2 in) thick polystyrene vertically along the walls from 30 cm (1 ft) above to 30 cm (1 ft) below the basement/wall joint. The exposed insulation 5 cm thick should be covered with flashing and siding. From the bottom of these panels insulation 5 cm thick should be placed along the ground 1.2 m (4 ft) out. The insulation should slope slightly down away from the house and be covered by 50 cm of soil and grass or walkways. The basement heat loss would then be controlled without affecting the summer cooling as discussed in Figure 10.

Excavation would be minimal and water would be carried well away from the basement, helping leakage problems.

WINDOWS AND THEIR INSULATION

Solar radiation through windows is an important source of heat. Some radiation is available in winter through east and west facing windows, but these will produce excessive heating in summer and produce a net winter heat loss as shown in the following table. South facing windows provide the maximum amount of passive solar heat in winter with an acceptably low summer heating load if suitable overhangs are used. For south facing windows double glazing is preferred if insulated drapes, or better still, insulated external shutters are provided. Double glazed windows transmit 72% of the solar radiation whereas a triple glazed window has only a 61% transmittance.

The following table shows the heat gain/loss through double glazed windows with drapes at night to improve the U-value. Clearly, north, east or west facing windows are undesirable from an energy standpoint.

Internal insulated drapes decrease the heat loss at night by preventing warm room air from reaching the inner window pane. The glass temperature will frequently fall below the dew point of the interior air resulting in condensation on the windows and dehumidification of the inside air. It is important for health and physical comfort to keep the humidity in the 30 to 40% range. Very low humidity excessively dries the skin and increases the transpiration loss from people which produces a sensation of being in a cold environment. If the humidity is kept up the inside temperature can be reduced two or three degrees for the same comfort. For this reason, external shutters,

TABLE: HEAT GAIN/LOSS THROUGH DOUBLE-GLAZED WINDOWS kW hr/m²

Daytime U-Value 3.349 W/m² K (0.59 BTU/hr ft² °F)
Nighttime U-Value 2.555 W/m² K (0.45 BTU/hr ft² °F)

MONTH	NORTH	EAST/WEST	SOUTH
January	−60.37	−48.95	+3.90
February	−48.57	−31.48	+17.53
March	−36.84	−11.02	+14.78
April	−12.09	+16.34	+9.39
May	+9.11	+37.43	+11.80
June	+26.66	+55.77	+23.45
July	+25.87	+60.11	+27.90
August	+18.26	+51.12	+41.62
September	+3.19	+28.04	+55.74
October	−13.59	+3.99	+53.82
November	−35.42	−26.68	+13.53
December	−54.61	−47.08	−5.56
June through August Totals	+70.79	+167.00	+92.96
October through April Totals	−261.94	−144.98	+107.40

Note: Positive values indicate heat flow into building, negative values out.

although more difficult to design, will be far superior to internal drapes. The glass surfaces are protected from the elements and hence reach room temperature and do not cause condensation.

A cross-section through an earth sheltered house is shown in Figure 12. Note the optimum placement of insulation as described in Figure 10, the roll type thermal shutters for the windows and summer shading by overhangs. Swing-type rigid thermal shutters could be designed to serve

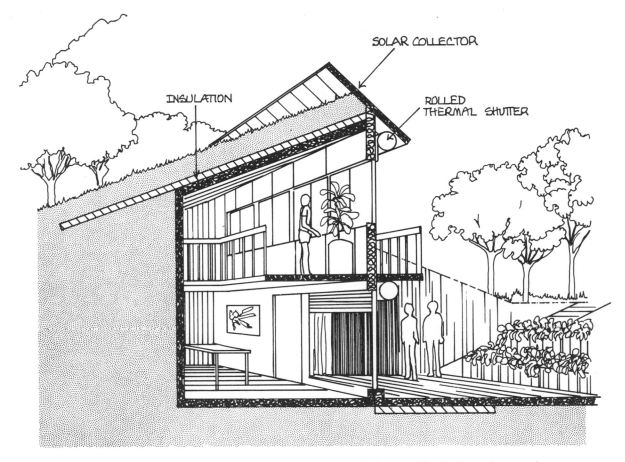

SOLAR COLLECTOR

INSULATION

ROLLED THERMAL SHUTTER

Figure 12. Earth-sheltered house cross-section showing correct placement of insulation and external window thermal shutters.

the dual purpose of winter insulation and summer shading. The whole area of window thermal shutters warrants more research and development since, per unit area, windows are by far the greatest source of heat loss.

REFERENCES

1. T.P. Bligh. "Energy Conservation by Building Underground," *Underground Space*, Vol. I, No. 1, 1976, pp. 19–33.

2. "Earth-Sheltered Housing Design — Guidelines, Examples, and References." The Underground Space Center, University of Minnesota, Minneapolis, 1978.

3. American Society of Heating, Refrigeration and Air-Conditioning Engineers. *Handbook of Fundamentals.* 1972.

4. E.R.G. Eckert, T.P. Bligh, E. Pfender. "Energy Exchange Between Earth-Sheltered Structures and the Surrounding Ground."

5. T. Kusuda. "Earth Temperature Beneath Five Different Surfaces." National Bureau of Standards, Report 10373. 1971.

APPENDIX 2

Financing Earth Sheltered Housing: Issues and Opportunities

MARK L. KORELL
Executive Assistant to the Federal Home Loan Bank Board

INTRODUCTION

Earth-Sheltered (E/S) dwellings are as old as humanity, when caves first supported human habitation. A similar concept was again used with success when American prairie settlers built their sod huts. Now in the late 1970's, the high costs and uncertain supplies of energy have stimulated a resurgence of interest in using earth as an insulator to improve the energy efficiency of new housing. While interest has been primarily individual rather than institutional, a scattering of pioneers around the country are constructing hundreds of prototype dwellings that can be aesthetically attractive while drastically reducing energy consumption. As interest grows, more and more questions are being raised with respect to financing

Presented at "Going Under to Stay on Top" Conference, Amherst, Massachusetts, June 8 & 9, 1979. Reprinted by permission.

problems and policies. How can adequate mortgage financing be made available in a helpful yet prudent manner? This overview attempts to raise a number of relevant issues and present some possible opportunities for action.

SPECIAL RISK FACTORS FOR FINANCIAL INSTITUTIONS

It is helpful and important to distinguish how lending on E/S dwellings may be different and more difficult than lending on more conventional homes.

Borrower default. One of the major risk factors in real estate lending is the possibility that the borrower may fall into arrears and default on monthly mortgage payment obligations. It is difficult to predict with much certainty where and when defaults will occur, because they tend to be correlated with economic conditions (job layoffs), personal circumstances (separation, divorces, death), and other factors. With E/S homes, would the likelihood of borrower default be greater or smaller? Because of the greater technical complexity of E/S dwellings, dysfunctional units could be constructed and would possibly result in more "walk-aways." On the other hand, it is argued that utility and maintenance costs will usually be significantly lower than normal, thus providing a "cushion" if economic adversity should strike. In addition, it is likely that E/S homes have generally been more attractive to better educated and more affluent households. If it is also true that well-designed and constructed E/S homes will cost more per square foot (which has often been the case because everyone is learning a new technology), then it is again likely that these homes will be most appealing to relatively higher income and more credit worthy borrowers. Note however, that full life-cycle costs of owning and maintaining the home over time may be substantially lower than average. Since this is a major unknown, how much "extra" investment would the prudent homebuyer invest in year one to gain savings over future years? To sum up, while there are additional factors involved, at this time it is unlikely that borrower defaults on E/S dwellings will differ from the norm.

Risk of loss on default. When a borrower does go into default, the investor/lender may or may not sustain a loss. If a lending institution foreclosed on the property, it would probably attempt to sell the property to recover its investment. The major issue here would clearly be: how widespread is the market for an E/S home and would the market price be consistent with or similar to the appraised value? The answer to this question involves a case by case situation, with important factors including price, location and surrounding environment, structural integrity, physical condition, and aesthetic qualities. A closely related issue would involve the level of original equity invested and the amount of "cushion" between the loan amount and the presumed market value. Clearly, the appraisal process itself is critical, and lack of experience and

market comparables creates substantial uncertainties.

Risk of loss can be significantly reduced through acquiring insurance, either from FHA/VA or private mortgage insurance companies. FHA has stated that E/S dwellings are 100% insurable through the normal 203B program, provided that Minimum Property Standards are met. For special experimental designs, the 233 option may be utilized and would require submission of plans to HUD/FHA in Washington. As a practical matter though, it should be noted that individual value judgments and lack of complete information can cause local variations from HUD/FHA's Washington policy and procedural guidelines. Although the maximum mortgage limit is $60,000, sales prices may be higher. Private mortgage insurance has, in fact, been written on E/S homes in a few cases and will generally cover the top 15–25% of the mortgage liability for the lender. While federal or private insurance would not necessarily be available for every type of E/S home, its utilization would not only reduce risk of loss, but could be helpful as a third party underwriting/appraisal "screen."

Liquidity of mortgages. Another factor bearing on the lender's risk picture is to what extent the investment can be sold in the

secondary market to another investor. The two largest such institutional investors are the Federal National Mortgage Association (FNMA) and the Federal Home Loan Mortgage Corporation (FHLMC). While a large private secondary market also exists, FNMA and FHLMC policies and procedures set the norm for secondary market transactions. If mortgages on E/S homes can be sold in whole or in part, the originating lender's risk can be virtually eliminated or at least reduced. Neither FNMA nor FHLMC have taken formal policy positions on E/S homes, and have concerns regarding the appraisal process and marketability. Since these organizations, in turn, sell securities in the national capital markets to finance their purchases, investment quality mortgages with minimal risk are required. In FNMA's case, FHA-insured loans would be purchased regardless of the type of property. For conventional mortgages, privately insured or not, an individual evaluation of marketability, the appraisal, location, and technical features would be required by either FNMA or FHLMC. Resale in the case of default is the big factor here.

Building codes. An additional area of possible risk to lenders involves building codes and standards. Where such state or local codes don't exist, the lender would have a difficult time determining structural soundness, safety, and even livability without rather extensive individual inspections. Where such codes do exist, while the E/S house may technically conform, it is very unlikely that the codes were written with E/S houses in mind. Thus, the special design and construction features that may very well be necessary would not be addressed by existing standards or codes, including FHA's Minimum Property Standards. On one hand, homes could meet all applicable standards and not be satisfactory; and on the other hand, the standards or codes may be overly restrictive or inappropriate so that perfectly adequate E/S homes would be prohibited from being constructed.

CURRENT PUBLIC POLICIES, ISSUES AND ACTIVITIES

Legislation/regulation. In financing E/S homes, there appear to be few, if any, prohibitions or serious constraints from the regulatory or legislative standpoint. In fact, the Congress and the state legislatures have rarely even recognized or addressed the subject. Federal regulators of financial institutions, namely the Federal Home Loan Bank Board, the Comptroller of the

Currency, the Federal Deposit Insurance Corporation, and the Federal Reserve Board, also have not explicitly looked into E/S structures nor prescribed any different criteria for mortgage investments. However, it is likely that federal or state examiners would want to look closely at any lender's substantial investment in E/S dwellings, because the unknown risks of loss could potentially weaken the financial soundness of the institution and, in a worst case scenario, force a merger or a bailout from the Federal Savings and Loan Insurance Corporation (FSLIC) or the Federal Deposit Insurance Corporation (FDIC). The bottom line is that each institution is free to make its own business judgment without any outright prohibitions or direction from statutes or regulations.

Research and demonstration. While an exhaustive search was not done, it appears that the State of Minnesota has launched the most substantial effort to date with respect to E/S housing. In 1977, $500,000 was appropriated by the State Legislature to the Minnesota Housing Finance Agency for the purposes of constructing 8–10 prototype homes and monitoring energy consumption characteristics. Several of the homes, which are scheduled for completion in 1979, will be built for State Park managers while others

will be sold on the open marketplace. An extensive design competition was held and response was heavy. The Underground Space Center at the University of Minnesota was awarded a contract to establish monitoring systems and publish results. Private financial institutions were also involved in a portion of the construction and permanent financing. This pilot program promises to yield valuable insights into many of the difficult design, construction, aesthetic, market and financing questions.

Federal Building Energy Performance Standards (BEPS). In what was billed as the first congressional intrusion into national building standards, Congress in 1976 passed the Energy Conservation Standards for New Buildings Act. The Departments of Housing and Urban Development and Energy (HUD and DOE) were instructed to develop standards for conserving energy through building design that states and localities were supposed to adopt in their building codes by early 1980. Although mandatory local adoption is not required and there is controversy over the substance of the BEPS, they should nonetheless raise considerable public consciousness regarding good energy design. This federal policy thrust could very well stimulate increased interest, research and credibility with respect to E/S concepts.

Other federal studies. The 1978 energy bill encouraged HUD to initiate research into financing and code practices with respect to E/S dwellings. HUD has contracted with Minnesota's Underground Space Center to prepare a report in 1979. DOE has likewise hired a consultant, DHR, Inc., to investigate both commercial and residential E/S structures. The study will include field research relating to financing obstacles as well as suggestions for appropriate federal roles in supporting and encouraging E/S construction.

It would be negligent not to mention solar energy in a discussion of E/S housing. Both HUD and DOE have initiated substantial efforts in the research and demonstration of solar heating and cooling technology. Because solar energy systems, active or passive, usually play an important role in E/S designs, the successful development of solar technology can clearly enhance the feasibility and acceptability of E/S housing. Solar and E/S can be viewed as independent but complementary energy-conserving technologies with major implications for dwellings of the future.

Underwriting and alternative mortgages. Lending underwriting standards are becoming more sensitive to energy and utility considerations. Many lenders and mortgage insurers are now looking at a home's energy usage and sometimes type of fuel as well. While this is generally now done in a rather informal subjective way, I would expect to see a more formal inclusion of utility costs along side of principal, interest, taxes, insurance and other debt. If the predictions are accurate that energy costs will rise faster than wages and income over the next several years, then these costs will have increasing potential to put homeowners in financial jeopardy and thus create greater risks for financial institutions.

Another related factor that has great potential significance is the changing array of mortgage instruments. For forty years, we have relied most heavily on the fixed-payment, self-amortizing mortgage. But with the rapid escalation of housing and financing costs over the past several years, many households have been priced out of the market. In fact, the National Association of Home Builders recently reported that only about one out of four households can afford to buy the average new home being built today. So what's being done about it?

• In California today, 225,000 mortgages on the books are a variable rate type, which means the interest rate, and therefore monthly payments, can rise and fall with economic and financial market conditions.

• About a quarter of all new FHA insured single family mortgages now contain a graduated payment provision where monthly payments start out lower than a conventional mortgage and rise by up to 7.5% a year.

• The Federal Home Loan Bank Board just last December approved regulations allowing federally chartered savings and loan associations to write conventional graduated payment mortgages nationwide, as well as variable rate mortgages in California only.

What this all means is that many more homebuyers will be entering into new types of mortgage commitments where they cannot count on a fixed monthly mortgage payment for 30 years. While these instruments have many advantages such as helping first time homebuyers and offering a choice for different family financial conditions, we also must be cognizant of the risks. A household's total monthly housing budget could rise more sharply with rapid increases in energy costs coupled with increases as well in PITI.

Therefore, in underwriting loans with non-fixed mortgage payments, lenders will undoubtedly want to look even more closely at the energy characteristics of the property. With E/S homes having the potential to reduce energy costs 30%, 50%, even 80%, they should enjoy special favor as a way to reduce the risk to lender and owner alike of monthly payment overload.

WHERE DO WE GO FROM HERE?

This final section will attempt to identify some opportunities for future actions that would reduce uncertainties and encourage further support in the financing of E/S homes.

Market experience. Perhaps the largest task that lies ahead is to overcome psychological barriers among both lenders and consumers. At this point, it would not be unusual to hear calls from Washington and elsewhere for a grandiose national study and demonstration funded by millions of tax dollars. While there is much merit in pilot programs to test new concepts, this topic of E/S dwellings lends itself nicely to more localized, small-scale efforts, like the one in Minnesota. To work out plans for a few homes with builders, architects, lenders, consumers, researchers and others allows the freedom to succeed, and to fail, on a manageable scale without the anticipatory glare of the national media, the bevy of bureaucrats and reporting forms in Washington, the political whims of the Congress. If you succeed in building that better mousetrap, plenty of attention will follow. All of the HUD or DOE studies in the world will not have the same impact as the building and sale of real homes in convincing both lenders and consumers to invest in E/S dwellings.

We simply need more real marketplace transactions between willing sellers and buyers. What is fair market value? Here again, reality may be much different from theory. To gain more acceptability and credibility, it is mandatory, in my opinion, to expand toward mainstream America. There will always be the wealthy and the eccentric, but lenders are not usually going to change their policies if this is the extent of interest. How broad is the market? Will there be a discount or a premium compared to conventional homes? These are business questions that need answers before too many lenders will risk having the federal or state regulators and examiners criticize them for taking imprudent gambles with their depositors' money. When all is said and done, consumer demand is the engine that will generate results. Unfortunately, we seem to face the classic chicken-and-egg dilemma.

Design and building standards. Current codes and standards need examining to determine where they are overly restrictive as well as where they do not provide enough guidance for E/S structures. The Federal government, with the aid of local experiences, could conceivably play a very helpful role in developing model standards for E/S dwellings and then disseminate them through state and local officials. If we continue to see an increasing number of do-it-yourselfers tackling all or part of building a house, then such standards become especially critical for guiding the design and construction as well as reassuring the lender. A few well-publicized blunders involving cave-ins, flooding or personal injuries could easily undermine and discredit the entire E/S concept.

While we know that architecturally pleasing and structurally sound units can and have been built, we don't know nearly enough about the relative importance and sensitivity of the design variables in real life. How much can we vary and in what ways from an ideal textbook house with respect to such items as waterproofing, soil thickness, foundation materials, stress loading and so on? How much and what parts of the construction can or should homeowners do themselves? More technical experimentation is definitely needed in order for lenders and investors to better understand the parameters of risk.

In reality there will be good E/S homes and ones that are not so good (just like conventional dwellings) and the trick will be to determine a methodology of judging the feasibility and soundness of technical and design features.

Real results. In order for the public to accept the E/S house for its energy conservation, sophisticated and credible analyses will have to be presented in addition to ad hoc individual testimonials. Widespread public acceptance of such a significant departure from the conventional dwelling as we know it, will require perhaps years of detailed monitoring of many units with different climates, designs, and settings. Government agencies, as well as universities and private researchers, will undoubtedly have to studiously document the energy-saving, lower maintenance, and other claims made on behalf of E/S dwellings. Put simply, the theories need broad real-life testing with actual numbers.

Secondary markets. Perhaps the fastest way to overcome lending institutions' fears of placing mortgages on E/S homes is to be able to readily sell the loans to another investor. As previously mentioned, FNMA

and FHLMC have cautious attitudes, but if just one secondary market loan purchaser would invest the research and analysis resources and then make a commitment to purchase certain types of loans, the local money would flow much more freely. Such action would set a standard, and shift risk to a much larger regional or national portfolio. The selling of participations in loans would also be an effective way to spread risk. Another related issue involves underwriting standards. A number of institutions, both originating and secondary, are moving toward more formal inclusion of a property's energy efficiency factors. This trend is expected to accelerate and will undoubtedly bode well for E/S dwellings.

Education. While the concept of using earth as an insulator in construction is old in theory and practice, the idea remains obscure and eccentric to most. With increasing amounts of small scale experimentation taking place, new data and living results will be available. It will need analysis, synthesis, and distribution. Some of the major educational efforts will need to be directed at the following groups:

- Architects who conceive, design and specify the plans.

- Public officials who sanction and inspect to protect the health, safety and well-being of the public.

- Contractors and labor who build the structures.

- Appraisers who estimate market value and acceptance, and investment risk.

- Lenders who provide construction and permanent financing and whose attitudes toward risk are traditionally conservative.

- Insurers who can reduce risk while a more extensive track record is being developed.

- Public policy makers who can thwart or accelerate the development of the E/S concept through legislation and regulation.

- Consumers who must make housing investment and lifestyle decisions.

In developing educational mechanisms, the federal government can certainly play an important catalytic role, but I would prefer to see more localized entities such as states and educational institutions take the lead roles.

Since cultural, environmental and climatic

conditions vary significantly throughout the country, it would seem appropriate to have regional resource centers whose sensitivity to local conditions and problems would enhance the efficiency and effectiveness of the educational effort. There is obviously room for many actors in this field, but in light of the general energy information efforts over the last 8 years, we now have the opportunity and the obligation to reduce overlap, duplication, and confusion.

Legislation. Now is also the time to look ahead to identify and anticipate legal impediments to broader scale utilization of E/S technology. This analysis does not involve such research, but one example could involve sun rights. Since most E/S designs incorporate active or passive solar heating (or cooling), assuring a home's legal access to sunlight vis-a-vis neighboring structures would be quite important in protecting the functional integrity and the value of the home. Other similar issues, especially relating to building codes, need further investigation.

CONCLUSION

With the modern day E/S technology in its infancy, new questions are constantly being raised. To begin to recognize and address some of the issues that bear on financing policies and procedures is of critical importance. For without constructive support from financial institutions, the growth and exciting potential of E/S structures will never be realized.

Heating Load Calculations
and Life-Cycle Costing Tables

<u>CONVENTIONAL HOUSES</u>

1. Data from solar study carried out by the University of Minnesota.
   Typical space heating load = 12,624 Btu/DD for 1200–1800 sq ft in Minnesota

   Minneapolis — 8382 degree days ↦ 106 million Btu per season (31,000 kW-hr)

   Thermal Integrity Factor assuming 1500 sq ft = <u>8.43</u> Btu/sq ft/HDD
   on living area

   = <u>4.22</u> Btu/sq ft/HDD
   including full basement

2. Data from RAND Corporation study. Average space heating load in west north central region of United States.

   <u>10.21</u> Btu/sq ft/HDD
   on living area

   <u>5.10</u> Btu/sq ft/HDD
   including full basement

These calculations and tables presented at "Going Under to Stay on Top" Conference, Amherst, Massachusetts, June 8–9, 1979. Reprinted by permission of Dr. Ray Sterling, Underground Space Center, Minneapolis, Minnesota.

DD = Degree Day